Light Trap
An Eco-friendly IPM Tool

The Authors

S.M. Vaishampayan, Retired in 2001 as Professor and Head, Department of Entomology, J. N. Agricultural University, Jabalpur. Obtained Ph.D. degree in Entomology in 1973 from the University of Illinois, U.S.A. Ph.D. thesis was related to visual and spectral specific responses of green house whitefly. In 1973 itself a systematic work was started on light trap studies at Jabalpur (M.P.). Work continued for more than 25 years till retirement in 2001, completing three research projects on light trap studies funded by ICAR, New Delhi. Technology was developed to use light traps as Pest Management tool more efficiently with improvement in trap designs and light sources etc. Served as member of ICAR Scientific Panel of Entomology 1984-87.

During the period 1978-80 he served as Forest Entomologist in Forest Research Institute, Dehradun, stationed at Jabalpur. Salient work included supervision of aerial spraying operations in teak forests, control of white grubs in teak nursery, work on Sal heartwood borer in outbreak years in the forest of Mandla and Amarkantak (M.P.). Returned back to J.N.K.V.V. in 1980 as Professor of Entomology and served as Head of Department for a period of 18 years. More than 75 research papers and review articles published. Research work continued after 10 years of retirement and developed new designs of light trap for MV and UV light sources in collaboration with Fine Trap (India).

Sanjay Vaishampayan joined J.N.K.V.V. Jabalpur in 1995 as Training Associate (Entomology) in Krishi Vigyan Kendra. Selected as Programme Coordinator K.V.K. in 2006. Recently posted as Senior Scientist (Entomology) at Directorate of Extension Services, J.N.K.V.V. Jabalpur. Obtained M.Sc.(Ag) degree in Entomology in 1989 from J.N.K.V.V. with gold medal and Ph.D. degree in Entomology in 1993 from Institute of Agricultural Sciences, B.H.U. Varanasi with first position in the subject. Ph. D. thesis work was on migration behavior of three lepidopterous pest species (*Heliothis armigera, Agrotis ipsilon* and *Plusia orichalcea*) analyzing light trap catches. Earlier, during 1994-95, worked at I.A.R.I. Regional Station, Karnal. Published 20 research papers in journals of repute and more than 120 technical bulletins in KVK.

Light Trap
An Eco-friendly IPM Tool

S.M. Vaishampayan

Retd. Professor and Head
Department of Entomology

Sanjay Vaishampayan

Senior Scientist (Entomology)
Directorate of Extension Services

JAWAHARLAL NEHRU KRISHI VISHWA VIDYALAYA
JABALPUR (M.P.)

2016

Daya Publishing House®
A Division of

Astral International Pvt. Ltd.
New Delhi – 110 002

Cataloging in Publication Data--DK
Courtesy: D.K. Agencies (P) Ltd. <docinfo@dkagencies.com>

Vaishampayan, S. M., author.
Light trap : an eco-friendly IPM tool / S.M. Vaishampayan, Sanjay Vaishampayan.
pages cm
Includes bibliographical references and index.
ISBN 978-93-5130-959-8 (International Edition)

1. Insect traps--India. 2. Insect traps. 3. Agricultural pests--Control--India. 4. Agricultural pests--Control. I. Vaishampayan, Sanjay, author. II. Title.

SB950.3.I5L54 2016   DDC 632.90284   23

Published by   :   **Daya Publishing House®**
                              A Division of
                              Astral International Pvt. Ltd.
                              – ISO 9001:2008 Certified Company –
                              4760-61/23, Ansari Road, Darya Ganj
                              New Delhi-110 002
                              Ph. 011-43549197, 23278134
                              E-mail: info@astralint.com
                              Website: www.astralint.com

Laser Typesetting   :   **Classic Computer Services,** Delhi - 110 035

Printed at   :   **Thomson Press India Limited**

डॉ० एस० एन० सुशील
Dr. S. N. SUSHIL
वनस्पति संरक्षण सलाहकार
Plant Protection Adviser

भारत सरकार
कृषि एवम् किसान कल्याण मंत्रालय
कृषि एवं सहकारिता विभाग
वनस्पति संरक्षण, संगरोध एवं संग्रह निदेशालय
एन.एच. IV, फरीदाबाद (हरियाणा) – 121001

Government of India
Ministry of Agriculture & Farmers Welfare
Department of Agriculture & Cooperation
Directorate of Plant Protection, Quarantine & Storage
N.H. IV, Faridabad (Haryana) - 121001
Tel. : 0129-2413985, 011-23385026
Email : ppa@nic.in

Foreword

Light trap as pest control tool was known to the entomologists and the farmers since early days, when insect control was based entirely on non-chemical methods. Around 1910-20 outstanding work was done on the control of red hairy caterpillars (*Amsacta* spp.) a serious pest of *Kharif* crops in the states of Haryana and Gujarat. Operations were carried out in campaign forms, using light traps in thousands of hectares of crop area in Hoshiarpur and Gurgaon Districts of Haryana. The pest (*Amsacta* sp.) was virtually eradicated from Govt. farm Nadiad, Gujarat within 5 years of operation (1911-16). It was a glorious past of successful use of light trap as pest control tool.

After 1940, however, with the invention of DDT, BHC and later organophosphates, entomologists and the farmers totally forgot to use light trap and other non chemical methods of pest control. For more than three decades, pest control was solely dependent on use of insecticides alone.

With the introduction of the concept of IPM and "Economic Threshold" and emphasis on non chemical approaches, use of light trap gained a wide spread importance in IPM strategies all over the world. Indian work was, however, very negligible in this respect. Exactly at this crucial stage, reviving the memories of golden past of early days. Dr. Vaishampayan and his associates initiated systematic work on light trap studies in 1973-74 at JNKVV, Jabalpur. Working consistently for over 25 years (with support of ICAR) they generated huge information and experimental data on various aspects of use of light trap in the pest management. Success story of eradication of *Amsacta* spp. was repeated almost 70 years later on Govt. farm Jobat (District Jhabua, M.P.) when pest was eradicated after 3 years of light trap operation (1982-84).

Finally, the authors made an attempt to publish a book on "eco-friendly, adult oriented technology of insect pest management without the use of insecticides" for its practical use to the farmers, and the scientists both. Most comprehensive and up-to-date on the subject, this publication would be of immense value to the people engaged in teaching, research and extension work related to control of insect pests in field crops, orchards etc.

Researchers of ICAR, SAUs *etc.* may carry forward research on light traps for effective monitoring and management of different group of insect pests in various crops of different agro-climatic regions of the country. The proven technology of light traps need to be disseminated amongst the farmers through various extension functionaries of State Govts., SAUs, Krishi Vigyan Kendras, Central IPM Centres, NGOs etc. all across the country.

I congratulate the authors for timely publication of this book which would help to create interest amongst the young scientists and research workers to take up the studies on light traps.

(S.N. Sushil)

Preface

Use of light trap in pest control was very popular in India in early days. Almost a century ago in 1910-30, work done by Indian scientist was just amazing and outstanding compared to the work done elsewhere in the world. Very successful control operations were carried out against *Amsacta moorei* in *Kharif* season and *Agrotis* sp. in Rabi season in a campaign form, using light traps in very large areas in the states of Punjab, Haryana, Gujarat and Bihar.

After 1940 however, with the discovery of DDT, BHC and other organic insecticides entomologists totally forgot to use light trap. For more than three decade's pest control was solely dependent on use of insecticides alone. It was an era of use, misuse and overuse of insecticides. This lead to the problems like pollution of entire eco system, health hazards to human beings and cattle, destruction of natural bio control agents (beneficial parasites/predators), development of resistance against insecticides and resurgence of pest population.

With the introduction of the concepts of IPM and "Economic Threshold" around 1973 and revival of non chemical methods of control, light trap gained a wide spread importance in IPM strategies in many parts of the world. Urgency was felt to use non chemical approach in pest control which is economically viable and environmentally safe. Use of light trap is one such approach where pest control is achieved without the use of insecticides.

In 1973, we initiated the systemic work on light trap studies at Jabalpur. With the active support of ICAR New Delhi, we carried out very extensive work on various aspects of light trap studies consistently for a period of almost 25 years. A technology was developed to use light trap as IPM tool more efficiently, suitable for scientific studies as well as for large scale use in farmers field.

Publication of a technical bulletin 'Summary of investigations of electric insect trap' (Edt: Hienton T.E.) published by USDA in 1974 was a milestone in this field, bringing a most comprehensive report on review of work done on a various aspects of light trap studies till that time and what further research is needed. We acknowledge with thanks the USDA for publishing such a nice report.

The book covers information on some theoretical aspects related to working of light trap like organs of insect vision and orientation towards light source, electromagnetic radiation, spectrum visible/attractive to insects, characteristics of electric lamps used, factors affecting trap catches and identification of migratory phases based on analysis of light trap catches etc. Practical aspects include principles and practice of light trap operation, trap designs and light sources to be used and study of seasonal activity and direct control of pest population using light traps. Last chapter covers extensive reviews of field trials and experimental work conducted on survey and control of some economic species of crop pests. The manuscript is supplemented in the end with a list of references cited in the text along with selected references on light trap studies for consultation to all the concerned.

The main objective of this book is to revive interest among the scientists, research scholars and students in using this novel technique of pest control and make it more popular among the farmers through research and extension work. We hope this book would serve as a reference/text book, meeting the essential requirements for advanced course 'Light trap as IPM tool' proposed to be introduced at Ph.D./M.Sc. level in all the agricultural Universities.

We express our gratitude to Shri Yogesh Paliwal, of Fine Trap (India), Yawatmal, (M.S.) for extending necessary facilities and encouragement in preparation of the manuscript of the book. Through mass production of light trap in plastic mould and their distribution to the farmers in various states in Indian Union he has fulfilled our dream to take the technology at the farmer's level on large scale. We are grateful to him for this contribution.

Sincere thanks to our wives Asha and Sakshi and the family members for their consistent support which enabled us to complete this assignment successfully.

S.M. Vaishampayan
Sanjay Vaishampayan

Contents

List of Figures

List of Tables

1
Introduction

Trapping of insects using light trap for Pest Control is an age old practice known to the entomologists in India since 19th century period itself when insect control was based only on non chemical methods of control including use of cultural and mechanical control practices. Light traps were very successfully used in pest control studies in Punjab, Haryana, Gujarat, Bihar and in parts of South India in early days.

Around 1910, a campaign was launched against the cutworms *Agrotis ipsilon* in Bihar using 55 Andres Maire Light traps. A total of 47523 moths were caught in traps in a season. An outstanding and unbelievable work was done during the period 1910-30 against 'Kutra' caterpillar (*Amsacta* spp.) covering large areas in a campaign form. At Government farm Nadiad (Gujarat) light traps were operated regularly for a period of 6 years during 1911-16. This resulted in declining the moth population, egg masses and larval population of 'Kutra' *Amsacta* spp. year after year and pest finally disappeared from the area after 1917 onwards.

As per report published in 1920, a large scale campaign was launched to eradicate hairy caterpillar pest (*Amsacta moorei*) in Hoshiarpur and Gurgaon districts of Punjab. In all 620 traps were installed in 21 different villages covering an area of over 3100 acres. In all, around 3,00,000 moths were trapped during 12 night's operation of light traps. During 1930 a successful campaign was carried out for the eradication of *Amsacta* spp. in Hoshiarpur district covering an area of 11000 acres using 2750 light traps capturing not less than 30 lakh moths in the traps.

After 1940 however, with the invention of insecticidal properties of DDT, BHC and later organophosphate insecticides and their availability to the farmers on large scale there was a sudden decline in carrying out such work. Farmers and scientists both totally forgot to use light trap and other non chemical methods in controlling the pest population in field. Instead, they relied on absolute use of these toxic insecticides indiscriminately and often in excess quantity. For every problem, irrespective of species or their feeding nature and density, use of insecticide application was the only answer. In a decade of 70ˢ and 80ˢ aerial spraying operations were carried out in large areas using these toxic insecticides against insect pests on paddy and cashew nut crops in the states of Kerala and Karnataka respectively. These organic insecticides, being highly persistent in nature, created several problems through pollution of entire ecosystem. Large number of cattle and human beings suffered with serious health hazards in the areas of aerial spraying. Besides the health hazards, over use of pesticides badly disturbed the natural ecosystem leading to destruction of natural enemies including beneficial parasites and predators acting as effective natural bio control agents. Besides this, the development of resistance in crop pest species against these insecticides was another major problem.

Entomologists all over the world were seriously concerned with these problems. Their serious deliberations and consistent efforts in the meetings held under the leadership of Entomological Society of America, USA in 1970 gave rise to a new concept of Integrated Pest Management and a supporting concept of Economic Threshold. These concepts emphasized on drastic cut in the use of toxic pesticides and utilize all the available non chemical methods of control to maintain the pest population below economic threshold level.

With the revival of interest in non chemical methods of control and more emphasis on ecological consideration, use of light trap gained a wide spread importance in IPM strategies all over the world. In the United States of America considerable work has been done on use of light trap in insect detection and control. Substantial progress was made in determining the potential and limitations of light trap for suppressing insect populations in a variety of crops. Vast amount of information existed in literature published in various journals on this subject. With the introduction of concept of IPM in 1970, U.S. Deptt. of Agriculture realized that there is a real need for someone to consolidate the available information on light trap studies and bring up to date information on what has been done to the present time and what further research is needed. The efforts finally lead to the publication of "**Summary of investigations of electric insect traps**" as a technical bulletin compiled by Trumen E. Hienton, the chief of the Farm Electrification Research branch, and published by Agricultural Research Station, USDA, Washington in 1974.

In 1973, Vaishampayan and his associates initiated systematic and well planned work on light trap studies at J.N. Agricultural University, Jabalpur (M.P.). With a consistent support of ICAR, New Delhi extensive work was done on various aspects of light trap consistently for a period of three decades "between 1973 to 2001". Major work included development of several designs of light traps (1982 to 2014) for efficient trapping, killing and preservations of insects in good conditions, seasonal activity studies of several major crop pest species recording daily collection of trap catches throughout the year, consistently for several years. Studies were made on influence of weather and moonlight on trap catches, testing efficacy of various light sources, studies on population dynamics and evidences on migratory behaviour of few species of major lepidopterous pests based on analysis of light trap catches. Significant work was done on the control of pest population in field against *Heliothis armigera* and *Agrotis ipsilon* in gram crop and *Spodoptera litura* in soybean crop. Valuable work was done on surveillance and control of red hairy caterpillar *Amsacta moorei* in the endemic area of Jhabua distt. (M.P.) near Gujarat border during the years 1983 to 1986. The pest was eradicated after three years of light trap operation on Govt. farm Jobat covering about 20ha. of area.

Work was seriously lacking on large scale use of light traps covering hundreds of hectares of crop area in a campaign form. Large area coverage on community basis in a campaign form is essential for true isolation of target area and effective control of pest population in field. Availability of light traps on large scale suitable for farmers use *i.e.* light weight and easy to transport maintaining its trapping efficiency, was another major problem. This problem has however, been solved in a recent past, when a young entrepreneur from Yawatmal Maharashtra, started production of light trap units in plastic moulds under the brand name Fine Trap (India), crossing the annual production of over one lakh units per year since 2010. On recommendation and approval of National Center for Integrated Pest Management (ICAR) New Delhi, these traps have been supplied to the farmers on subsidized rates through Deptt. of Agriculture in various states of India including states of North East region. This trap was designed by NCIPM and claimed to be eco-friendly, being safer to beneficial parasitic and predatory insect species acting as bio control agents. The authors however, are of the opinion that no light trap can be safer to natural enemies by virtue of design. Use of light trap itself is an eco-friendly approach trapping adults of insects which are active during night hours only. Most of such species, being highly phototrophic in nature are selectively attracted towards the light source of the trap in large numbers. Contrary to this fact, most parasitic species belonging to the orders Hymenoptera and Diptera and few of the predatory species being diurnal in habit are active during daylight only. Therefore, population of such species remains undisturbed by light trap operation. In observations made at Jabalpur, the proportion of such beneficial species was observed to be only one percent compared to total biomass of insects collected in the trap.

Light trap is one of the effective tools of management of the insect pests as it mass-traps both the sexes of insect pests and also substantially reduces the carryover of pest population in the following season. It's regular operation for few hours every night in major active period of the pest species in field helps to control the pest population and minimize economic losses to the crop significantly without the use of any insecticides. The key insect pests of cereal crops (rice, maize, sorghum), pulse crops (chickpea, pigeon pea), soybean, sugarcane, vegetable crops (okra, cauliflower, cabbage, tomato, brinjal) and horticultural crops (apple, mango, pomegranate, oranges etc.) can be mass trapped in adult stage by using the light trap. The major insect pests that are attracted towards light trap include the rice leaf folder, rice stem borer, gundhi bugs, gall midge and leaf hoppers, stem borers in sugarcane, codling moth, cabbage looper, cutworms, armyworms, pod borers, diamond back moth, webworm moths, leaf roller moths, tobacco caterpillar, bark beetles, red hairy caterpillar, white grubs, groundnut leaf minor etc.

With the exception of significant work done at JNKVV Jabalpur and few other centers including research institutes and All India coordinated research projects of ICAR, there is a lack of awareness in general, among the Indian scientists and research workers regarding utility of light trap as a pest control tool. Most of them are unaware of the importance of proper design of traps and light source needed for effective control of the pest population. With these lacunae in mind efforts have been made to consolidate all the information available on light trap studies and compile it as a reference book for use to all the concerned. This book may serve as a 'Textbook' for a 3 credit course work on "Light trap as pest management tool" at M.Sc. or Ph.D. level in the Agricultural Universities. One of the aims of writing this book is to create interest among the research workers and graduate students working in various Agricultural Universities/colleges and ICAR institutes to initiate some work in light trap studies of their interest*.

Starting with the early history of light trap including the Indian work the contents cover information on various theoretical and practical aspects related to operation of light trap. Theoretical aspects include structure and functioning of insect vision, mechanism of orientation of insects towards light source, physical aspects of electromagnetic radiation and range of its spectrum visible to human and insect eyes. A separate topic covers information on various types of electric lamps used and their spectral characteristics with reference to the response of insects attracted toward light trap. Practical aspects include factors affecting light trap catches, designs of traps and light sources used. A chapter is devoted on principles and practical use of light traps in monitoring seasonal activity of pest species and direct control of the pest population in field, supported by presentation of experimental data collected on different pest species.

Interesting information is compiled in a chapter on 'Role of light traps in a study of insect migration' with specific reference to identification of migratory phases of population based on analysis of related morphological and physiological characters of the adults collected in the light trap. Separate chapter is included on review of world literature, including Indian work, related to field trials conducted on survey and control of insect pest species. It further includes review of experimental work done on testing of various light sources as an attractant against many species of crop pests conducted under laboratory and field condition both.

Future Line of Research

Further research work on light trap studies needs to be undertaken against the potential pest species on following three major aspects.

1. **Seasonal activity studies of major crop pest species collected in light traps.** For seasonal activity studies at least two light traps be installed around 100 mt. away at each location. Operate the trap every night using 8 to10 w UV lamp (two tubes) or 80 w MV lamp. Data of daily trap catches may be recorded species wise to study their activity on weekly, monthly or yearly basis as needed. Mercury vapour is a good source of light for majority of species. But owing to its high electricity consumption,

---

* in support of this adult oriented novel technology of pest control.

requirement of heavy weight choke and some operational problems in field it needs to be replaced by black light 15 to 20 w UV lamps.

2. **Evaluation of relative performance of various potential light sources *i.e.* UV, MV and LED lamps for their use in light trap.**

   For evaluation of performance of various light sources at least two traps should be installed. This will help to study the performance in a paired tests comparing UV (10+10)w or 15w as principal source with MV and greenish yellow LED lamps. There is an urgent need to generate research data on practical use of LED light sources, available with emission in 5nm bands in a range of 360nm to 600nm wavelength (ultra violet, violet, blue or greenish yellow radiations).

   Information is needed on type of light source most attractive to insects of different groups of orders specifically the order Lepidoptera, Coleoptera and Orthoptera in one group and all the Hemipterous and Homopterous insects including various leafhoppers specially the BPH and *Pyrilla*, gundhi bugs, whiteflies and some dipterous flies like gall midge etc. in other group. Greenish yellow light with radiation in visible range of 520 to 580nm wavelength is highly attractive to insects belonging to this group.

3. **Studies on use of light trap as direct pest control method.** Effect of mass trapping of adults on suppression of pest population in field by operating the traps everyday in the active season needs to be studied in two ways (i) testing in individual field using one or two traps (ii) testing on area wide basis by installing the traps in large areas on cooperative basis, covering about 500 to 1000 ha of crop area in a single village. Following standard procedure of sampling insect population observations can be recorded on level of damaging stage of insect *i.e.* larval, nymphal or adult stage per sample unit, in trap area v/s non trap area (control).

   There is a need to take up such experimental work against codling moths (*Cydia pomolina*) in apple orchards and godowns and against *Agrotis* sp. (cutworms), *Spodoptera litura* and white grubs in vegetables and tuber crops in North Western Himalayan region including Leh, Laddakh, Kashmir etc. where organic farming is most common and use of insecticides is strictly prohibited. Large scale adoption on community basis is required for effective pest management. Such trials are also needed to be undertaken in a campaign form at least against two major species *i.e.* hairy caterpillar *Amsacta moorei* and black cutworm *Agrotis ipsilon* in the endemic areas of different states as was done 100 years ago in the states of Punjab, Gujarat and Bihar respectively.

2
Early History of Light Trap

The attraction of some adult insects to artificial light was observed and recorded by man early in his history. Fires undoubtedly provided the first artificial light to attract and destroy flying insects. Harris (1821) indicated that small fires were desirable "as a mode of destroying insects." The use of fires for luring and killing adults of the armyworm *Pseudaletia unipuncta* (Haworth); cutworm (Noctuidae), and codling moth, *Laspeyresia* (=*Carpocapsa*) *pomonella* (Linnaeus) was mentioned by Glover (1865).

Developments of Trap Models at Global Level

Kerosene-burning lamps and lanterns succeeded fires as insect attractants. These light sources were most frequently incorporated into trapping device known as a trap lantern. Knaggs (1866) reported the invention of "Glover's new American moth trap". Mr. Glover was the first entomologist of U.S. Deptt. of Agriculture. This trap used a kerosene lamp as the attractant inside a box with an entrance of glass sheets opposite the lamp (Figure 1). It was listed by Wilkinson (1969) as the first well designed entomological light trap and its early use and later modifications were described in detail by him.

The development of 11 trap-lanterns of widely different designs, 10 of which were patented, occurred in the United States during the period 1864-77. In 1878, this equipment was tried on a large scale. More than 1,000 lanterns were used near Hearne, Tex. to capture moths of the cotton leaf worm and the bollworm, Comstock (1879).

Riley (1885) observed the attraction of cotton insect moths to electric lights around an Atlanta Hotel in 1881. The first electric light trap with funnel and baffle, known to the author was that reported by McNeill (1889) *designed,* constructed, and operated in 1888. He used a funnel 6½ inches in diameter, with a single tin or glass

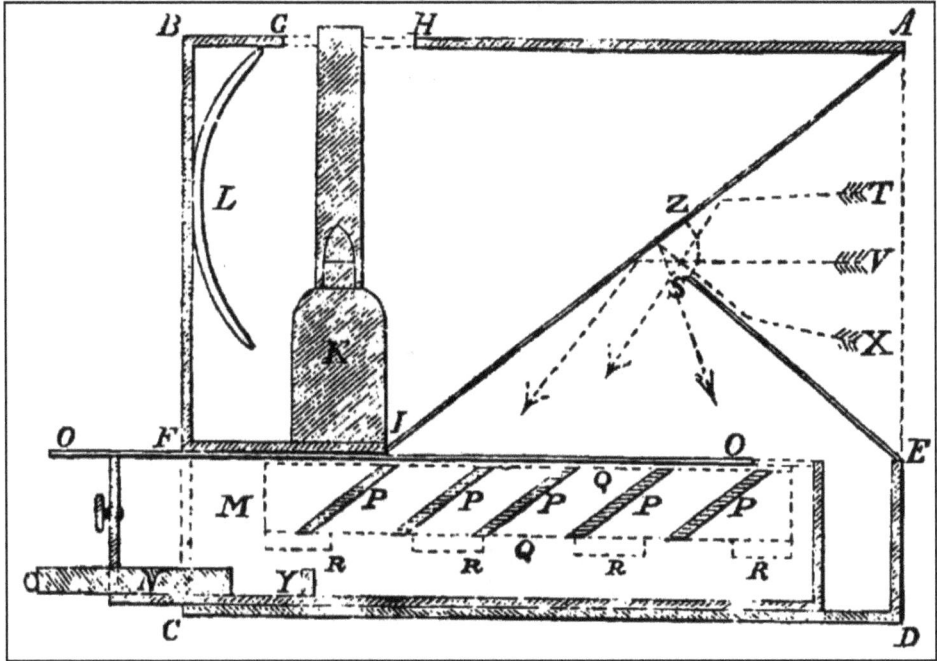

Figure 1: Glover's New American Moth Trap from Knaggs (1866).

A B C D is a box, having a partition, F for lamp K to rest on, behind the latter being a strong reflector L. The box is open at A E, also at H G (for lamp chimney to pass through), and at F C for the drawer M. M is a drawer fitted above with a glass slide O running in a groove. It is also fitted with a small drawer N, which is filled with layers of flannel for the reception of chloroform, and stopped by the block Y.

baffle that stood vertically across the center of the funnel rim (Figure 2). Slingerland (1902) made extensive studies in New York on the kinds of insects caught in light trap lanterns in 1889 and 1892.

The development of several trap-lanterns in France using acetylene was reported by Vermorel (1902). In Montana, a trap was made up of utensils commonly found on farms and which served other purposes when not in use as a light trap (Parker, and others 1921). The trap consisted of galvanized-iron washtub and barn lantern. A galvanized-iron arch was fitted across the tub and served to deflect the moths and to hold the lantern (Figure 3). Eleven such traps caught 82,488 pale western cutworm moths, *Agrotis* (=*Porosagrotis*) *orthogonia* Morrison, during the 1920 season.

One of the first recorded insect trap applications using the incandescent electric lamp, was that by Runner (1917). He enclosed the lamp with sticky fly paper to trap attracted cigarette (tobacco) beetles, *Lasioderma serricorne* (Fabricius). He noted that the adult beetles flew "more readily to blue or violet light than to red or orange." Studies of insect catches in orchards by Parrott (1927) and of the Oriental fruit moth, *Grapholitha molesta* (Busck), by Peterson and Haeussler (1926) were made with traps in which pans, filled with water and a film of oil, were fastened below an

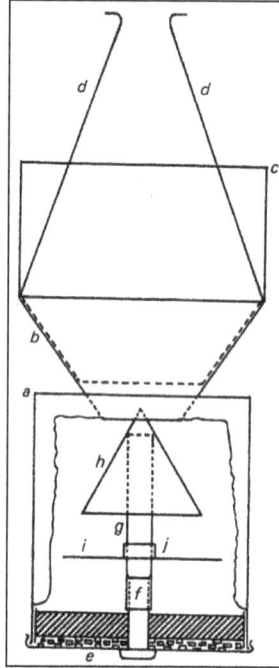

Figure 2: McNeill (1889) Insect Trap to be Used with Electric Light.
a, 3-quart tin pail; b, funnel; c, baffle, soldered across center of funnel; *d,* steel wires support trap; *e,* pail lid forms trap bottom; f, tube; g, cylindrical tube; h, hollow cone; and i, disk.

Figure 3: Montana Light Trap (1919) Parker *et al.* (1921).

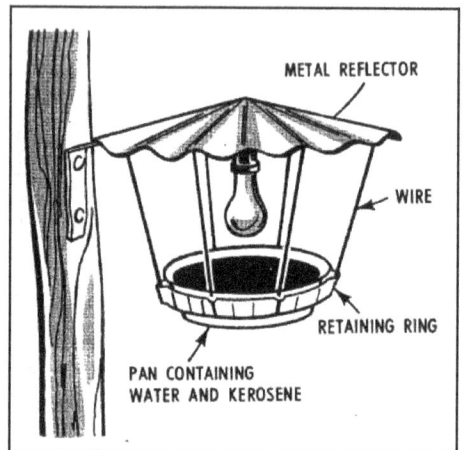

Figure 4: An Early Light Trap Used in Virginia and Alabama (1928).

electric lamp. Farmers in Virginia and Alabama used similar trap in1928 (referred by Hienton (1974) to control the tomato fruit worm (Figure 4). The trap was very simple, consisting of a reflector, lamp, and water pan Electric insect traps used for *Cyclocephela* in 1939 with a common metal wash tub filled with water which served as the insect retainer is shown in Figure 5.

Figure 5: Electric Insect Trap (1939).

Early History of Indian Work on Light Trap

In early days, before the invention of insecticidal properties of DDT and BHC in 1940, light traps were very commonly used by the entomologists in pest control studies. Almost 100 years ago very extensive and outstanding work was done on use of light trap against *Amsacta* spp. and *Agrotis* spp. in North India

The earliest Indian record of use of Light trap in pest control is that of Dutt (1919) who carried extensive work on control of black cut worm in Bihar. A campaign was launched against *Agrotis ipsilon* at Ghoza and Colgong in Bihar with 55 Andres Maire traps. A total 47,523 moths were caught in the traps during the season of which 22889 were males and the rest females. In all, 587084 caterpillars were destroyed. The details of Andres Maire traps are however, not known.

Jhaveri (1921) reported that the use of light traps over a period of 6 years on the Government farm Nadiad (Gujarat), during the years 1911 to 1916 resulted in declining the moth population, egg masses and larval population of Kutra (*Amsacta* spp.) year after year and pest finally disappeared from the area after 1917 onwards.

Milne (1927) advocated the use of light traps for the control of *Amsacta* spp. in Punjab, doing extensive damage to maize, jowar and sunhemp etc.

Chopra (1928) also reported success in use of light trap in eradication work done against kutra hairy caterpillars *Amsacta* spp. in Hoshiarpur and Gurgaon districts. In all, 500 light traps were installed in 21 different villages spread in 4 Tehsils, over an area of 2500 acres, operated for a total period of 15 nights in Hoshiarpur. At Gurgaon, 120 light traps were set up and operated for 12 nights over an area of 600 acres in different localities. Some 3,00,000 Kutra moths were trapped in both the districts. In these areas almost all the *kharif* crops grown were free from kutra damage compared to those fields where no eradication works were carried out.

Hussain (1930) reported results of successful campaigns against the Kutra (*Amsacta* spp.) in Hoshiarpur district using light trap at widely separated localities covering a total area of 11,000 acres. About 2750 light traps were set up capturing not less than 30 lakhs moths.

Dina Nath (1923) made extensive observations in the sugarcane fields and recorded 3194 moths of three species of cane borers (*Chilo simplex 2180* moths,

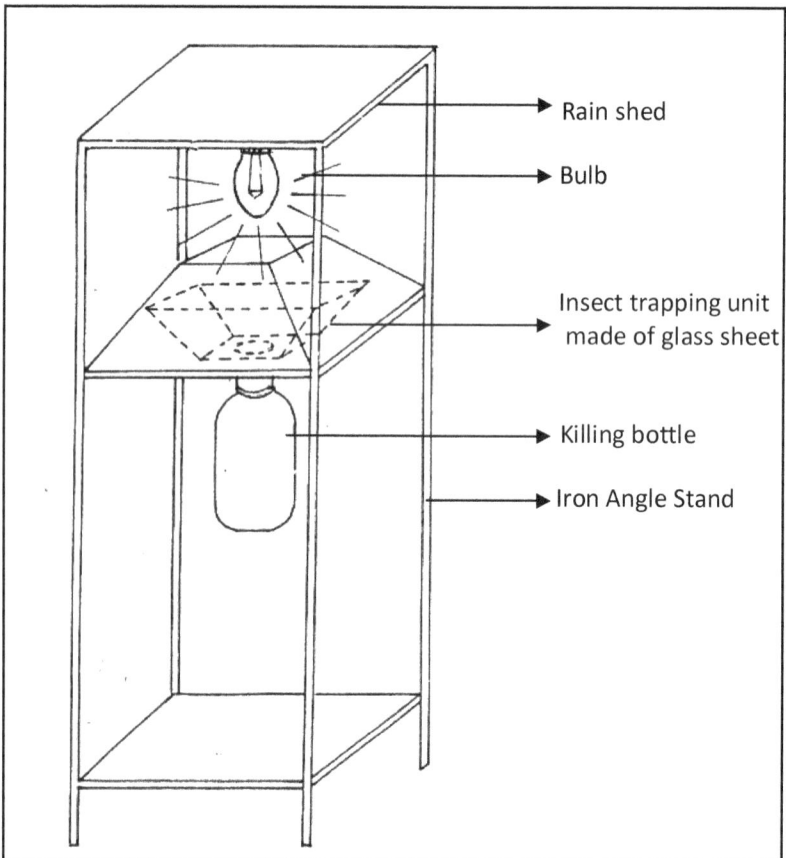

Figure 6: Chinsura Trap (Banerjee and Basu, 1956).

Scirpophaga sp. 451 moths and *Emmalocera* sp. 563 moths) during 59 nights of trap operation.

During 1920-30 light traps were listed as an effective method for the control of many pest of economic importance (Hussain 1934 and Anathanarayan, 1937).

As reported by Ayyar and Ananthnarayanan (1935 and 1937) light trap proved very useful in minimizing infestation of Rice stem borer *Schoenobius incertulas* in South India. They were able to collect 15427 moths in 29 nights.

Banerjee (1949) initiated the work on the study of rice yellow stem borer *Tryporyza incertulas* (Walker) activity in relation to weather factors with the help of light trap. A new design of light trap named as Chinsura Trap was developed by Banerjee and Basu (1956), suitable for pest surveillance studies. The trap was provided with proper trapping device, made of clear glass sheet and a glass jar for collection of insects (Figure 6). The trap was extensively used in pest control studies in West Bengal and many other parts of the country.

In Fourth Plan (1973-78) Dr. S.N. Banerjee, as Head of Directorate of Plant Protection, Quarantine and Storage, Govt. of India introduced light trap work on all India basis under the All India Surveillance Programme. Work was started on 19 centers operating 6 light traps at each center from sunset to sunrise every night (Banerjee, 1985).

3

Structure and Functioning of Organs of Insect Vision

Structure of Compound Eyes

Insect's responses to the surrounding world are determined to great extent by visual stimulants. Through their eyes insects receive information which quite often man is unable to receive. Among the higher insects the Pterygota, compound eyes occur in almost all adult individuals. Compound eyes are faceted eyes or paired organs of vision, placed on the either side of head and intimately connected to the visual centers of the brain, which are usually highly developed. Their size, shape and internal organization vary greatly among various taxonomic groups. As a rule, the mobile and fast flying species, both diurnal and nocturnal, have large complex eyes. Insects living a hidden and sedentary life have small eyes. In most insects the faceted eyes are rather voluminous compared with the size of the head. The largest eyes are found in dragon flies and Dipterans flies.

The eye's surface is formed by a transparent, chitinous cornea, which acts like a multiple lenses made up of minute, regular hexagonal protuberances called facets; hence, the name faceted eyes. Each facet is the corneal lens of a separate functional unit of a vision called ommatidium and the array of ommatidia forms the compound eyes. The corneal lens is transparent to ultra violet.

The structure of ommatidia, in typical unit, is made up of an array of photo refracting (dioptic) photosensitive and photo insulating (pigmental screening) elements. Dioptic elements consist of two basic elements (i) the corneal lens and (ii) crystalline cone, which in most cases function as an optical unit.

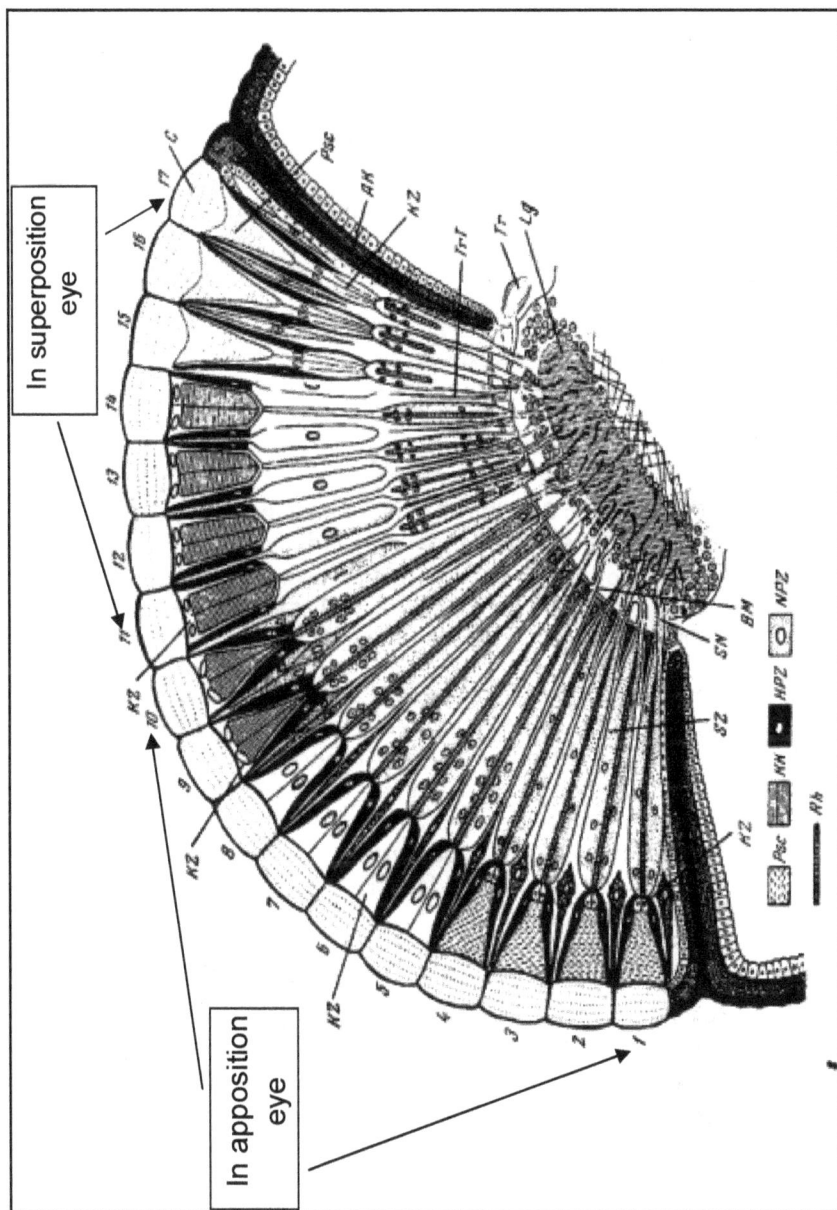

Figure 7: Schematic Structure of Faceted Compound Eye After Weber (1949).

1-10: ommatidia of an apposition eye; 11-17: ommatidia of a superposition eye; C: cornea; KK: crystalline cone.

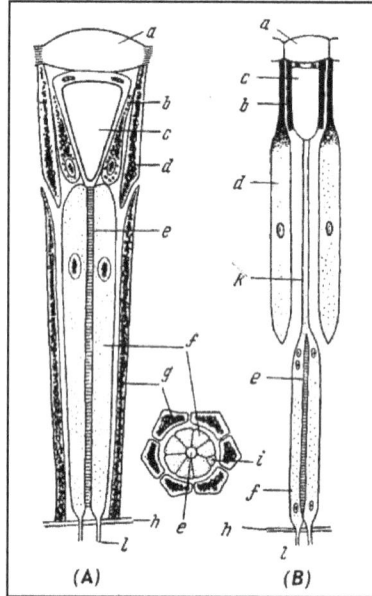

Figure 8: Schematic Structure of the Ommatidium of an Apposition (A) and Superposition (B) Eye.

a: Cornea, b: Main (corneagenous) pigment cells, c: Crystalline cone, d: Pigment cells, e: Rhabdom, f: Visual cells, g: Retinal pigment cell, h: Basilar membrane. In the middle: cross-section of an apposition eye through the rhabdom.

The designation of eyes as apposition type (suitable to function in day light) and superposition type (for vision in weak light) were proposed by Exner in 1891 and still hold true at present. Superposition eyes have a short rhabdom and a variable light insulation of the retinula. The iris cell pigments can shift toward the distal end so that light can reach rhabdomere even through neighboring facets. Superposition eyes are organs of vision characteristic of nocturnal insects, but they are possessed by some diurnal species also such as Hesperiidae, Macroglossum, Anarta etc. closely related to typical nocturnal species.

Visual Stimuli as a Means of Recognition and Orientation

The visual recognition of objects and the corresponding behavioral reaction of the organism are realized through various visual stimuli perceived by eyes. Light is an important factor which mediates all these responses. The information is transmitted to the insect's eyes through changes in colour or spectral composition, intensity, plane of polarization and temporal properties (motion of flashing) of light. All these parameters may be present in a visual stimulus and their perception is necessary for the realization of all visual reactions. Insect's behavioural response is the final act of chain of reaction process beginning with perception of light by the eyes.

Insects make use of light as an index and signal of the environmental conditions as open space is directly illuminated by the light from the sky and it is always brighter than a secluded space. Weak illumination is a sign of confined space, a space for shelter and protection. On the contrary bright light is the sign of an opening from the shelter leading towards the open space outside. If it needs to, the insect now will move toward bright light. This is the point of phototaxis.

An animal reacts to different objects and situations according to the information the sensory mechanism receives through light of a given wavelength and its influence on the effectors (motor) mechanism. In certain species of insects, in particular environmental situations, the visual stimulus act as 'super normal sign stimulus' and evokes a strong specific reflexes (fixed action pattern) leading to their orientation towards a light source (positive phototaxis) or away from the source (negative phototaxis). These reactions are guided by the type of signals the visual stimuli provide to the motor centers and may be related to the (1) feeding, sexual or defensive reflexes or (ii) the habitat conditions.

(a) Reactions to Feeding, Sexual or Defensive Reflexes

Some butterflies are attracted by blue colour because they are accustomed to feed on blue coloured flowers; to them blue is a colour of food. For other insects orange is the colour signifying female, for still others green is the colour signifying protection, a background against which the insect is less visible to enemies.

(b) Indication of Habitat Conditions

By choosing between bright and dim light or between light and darkness insects make use of light as an index and signal of the environmental conditions, whether the surrounding space is open or secluded.

When we consider that light attracts insects because it is a sign of open space, then it appears that UV radiation delivers the most typical indication of open space. By day sun and sky are the main sources of ultraviolet, violet and blue radiations, which are diffused by the atmosphere. By night, or when the Sun is not visible, the sky is the main source of shortwave radiations. The light falling from the sky is a sign of spaciousness. When insects need open spaces they direct themselves toward the light, when they want to hide, they run away from it. This is why a lamp will attract the nocturnal insects from a greater distance when the emitted light is similar to the natural nocturnal light, only it is brighter (Mazokhin- Porshnyakov, 1960).

Mechanism of Insect's Orientation Towards Light Trap

Various theories have been proposed to explain why and how insects fly toward light and are discussed extensively by Mazokhin Porshnyakov (1969), Hardwick (1968) and Southwood (1978). The exact mechanisms that lead to insect's capture by a light trap are still far from clear.

According to Buddenbrock theory – insects fly to a light source along a logarithmic spiral, maintaining the constant angle between the axis of its body and the direction of the incident light. This is a case of transverse orientation. This theory has, not been confirmed.

Marten (1956) proposed another hypothesis. He believes that insects fly to a light source, taking it as exit from a confided space beyond whose limits they will find open space and unlimited freedom of flight. If a noctuid moth is placed in a room by day it will immediately fly to a dark corner. A diurnal butterfly, instead, behaves in exactly the opposite way. It flies straight to the window and crawls onto the window panes trying to escape to freedom. As evening approaches their behavior reverses. Thus, light attracts both diurnal and nocturnal, one by day time and other by night, evoking in them positive phototaxis because to them it is a sign of 'escape to freedom'.

Robinson and Robinson (1950) and Robinson (1952) suggested another, almost diametrically opposed theory of the light response phenomenon. They suggested that nocturnal insects are actually repelled by light, but that if they are flying sufficiently quickly, they may come close enough to the light source to be 'dazzled' by it and are then automatically deflected toward it.

According to Verheijen (1960), the light trap catches insects because the light illumination of trap relative to the surrounding interferes with the normal photic orientation, this results in insects moving towards the light source. Unusual light distribution disturbs the normal co-ordination of motion, and the animal deviates from its initial course. In terms of orientation mechanism described by Fraenkel and Gunn (1961) it would be a case of tropotaxis.

In conclusion, the hypotheses proposed by Marten (1956), Verheijen (1960) and Mazokhin Porshnyakov (1960 and 1969) seem to be more logical and perhaps, in combine, explain the phenomenon of attraction of insects and their trapping in light trap.

Mazokhin Porshnyakov (1969) explained the behavior of positive phototaxis as follows: - Insects do not fly away from lamp because when they get into the region of high luminous flux, even for a short time, their eyes rapidly become light adapted. Ultraviolet and long wave radiations cause migration of the ommatidial protective pigments from nocturnal to a diurnal position. Thus, the visual sensitivity decrease by hundreds and thousands of times (Bernhard and Ottoson 1960 and 1962) and the insects cannot distinguish objects. The lamp is the only thing that insects are now able to see, and they cannot fly to the sides of it in the darkness of light.

Thus, insect's attraction to light and their trapping is purely a function of contrast effect in illumination. The phenomenon of trapping very fewer insects in light trap on full moon nights (lesser contract effect) compared to dark nights around no moon can be explained on the same basis. Another explanation is of observation of higher efficiency of light trap catches in southern regions where nights are darker than it is in northern region where contrast between artificial and natural illumination is low.

Phototropic Response of the Codling Moth and Physiology of the Compound Eyes (Tashiro, 1961)

In ordinary behavior the moths are at rest during day. As evening approaches, they fly around the upper leaves, engaged in mating flights and in oviposition. Two

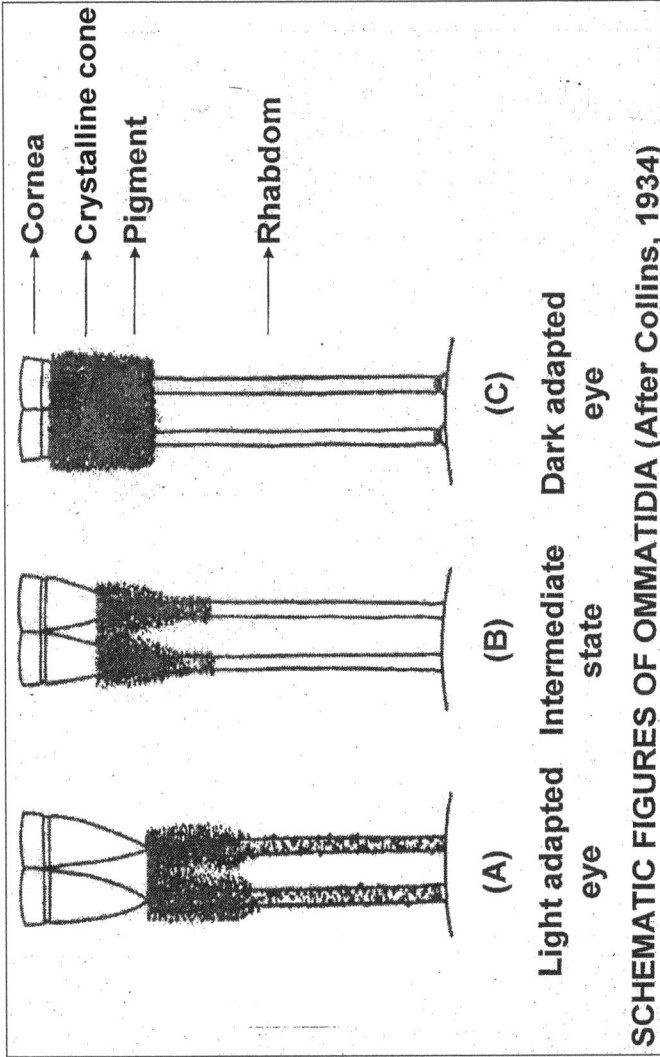

Figure 9: The Distribution of Protective Pigment in the Superposition Eye of Codling Moth (*Carpocapsa pomonella*) as a Function of the Adaptation State of the Eyes.

SCHEMATIC FIGURES OF OMMATIDIA (After Collins, 1934)

(A) Light adapted eye
(B) Intermediate state
(C) Dark adapted eye

Cornea
Crystalline cone
Pigment
Rhabdom

periods of inactivity were found to be due to the adaptation of the compound eyes. During the day they are said to be in the light adapted condition and during night they are in the dark adapted condition (refer Figure 9).

The iris pigment migration was found chiefly responsible for moths being in these two states. In the light adapted condition the iris pigment is withdrawn from the area of the crystalline cones and migrates toward the apices of the retinulae cells. Pigment distribution thus prevents sufficient light from reaching the retina to cause a phototropic response. In the dark adapted condition there is a dense mantle of iris pigment around each crystalline cone and withdrawal of the retinulae pigment towards the basement membrane. Dark adaptation did not begin until about 15 minutes before sunset and was completed 30 to 50 minutes after sunset. Migration towards light adaptation started about 30 minutes before sunrise and was completed.

What Insects are Positively Photosensitive

As reviewed by Oman 1961, insect vision is basic to orientation, movement and consequently insect environment. There appears to be no simple, uncomplicated answer to the question of what insects are attracted to induced light. Whether or not an insect exhibits a positive response depends upon various circumstances, some of which concern the insect itself, some of which depend upon its environment, and some that depend upon the nature of the induced light. There are thousands of kinds that respond to light, in varying degrees, under certain favorable circumstances.

The bulk of the photopositive species of insects belong to the orders Ephemeroptera, Neuroptera, Orthoptera, Hemiptera, Coleoptera, Trichoptera, Lepidoptera, Diptera, and Hymenoptera. In general, diurnal species are not attracted to induced light. The great majorities of the species that show a marked positive response to light are nocturnal, or carry on some function essential to species survival at night or at dusk or dawn. Examples of these essential activities are emergence to the adult stage, mating, oviposition, feeding etc.

Well-known pest species among the Lepidoptera, the adults of which may be attracted to light, are the codling moth, oriental fruit moth, corn earworm, various cutworms, fall armyworm, red hairy caterpillar, cabbage looper, European corn borer, pink bollworm, and others. Many respond only to certain wavelengths of light, and most show definite time peaks of nocturnal activity. Lepidopterous families containing numerous photopositive species are Noctuidae, Notodontidae, Sphingidae, Arctiidae, and Geometridae, as well as many Microlepidoptera of various families.

The Scarabaeidae, Staphylinidae and Cerambycidae among the beetles, contain many species that are attracted to light. However, the Japanese beetle, a diurnal scarab, is not so attracted. The Asiatic garden beetle and the European chafer and few species of white grubs (*Holotrichia* spp.) are strongly attracted to light.

In the Diptera most of the positively photosensitive species occur in the Nematocera, particularly in the Chironomidae, Ceratopogonidae, and Culicidae. Among the Brachycera, members of the family Pyrogotidae and the tribe Ormiini of the Tachinidae are attracted to lights. Other photopositive Brachycera occur in the Empidae, Lonchopteridae, and Sphaeroceridae.

In the Hemiptera the sternorrhynchous Homoptera are little attracted by induced light, although aphids may be more strongly photopositive than has been suspected. Of the auchenorrhynchous Homoptera the leafhoppers and fulgoroids, particularly the former, contain many species that are positively photosensitive. Among the terrestrial Heteroptera, several lygaeids, mirids, and cydnids are strongly attracted to light, as are most corixids and some other aquatic species.

4

Physical Aspects of Electromagnetic Radiation, Light and Colour

Electromagnetic energy is only a one form of energy known today. Other forms are thermal, chemical, atomic, electrical, etc. Electromagnetic energy is referred to as radiant energy because it exists in the form of repeating wave patterns. ''Radiant Energy Attractants'' are those sources of electromagnetic radiation which are capable of inducing insect to move towards and closely approach the source of radiation.

The Electromagnetic Spectrum

Radiant energy is released in bundles of quanta travelling in waves of different lengths and heights but at the same speed. All radiant energy travels through free space with the velocity of light at the speed of 186,000 miles per second. The electromagnetic spectrum is made of several kinds of radiation. All these radiations *i.e.* radio frequency, infrared visible, ultra violet, x ray, gamma (Y) and cosmic rays, etc. differ from one another only in wavelength and related characteristics. At one end of the spectrum are cosmic rays, and at the opposite end are electrical power waves. The average wavelength of the shortest cycles of radiant energy known as cosmic rays is 0.00001 (nm). At the other end of the spectrum is electric power waves- average wavelength of almost 5 million meter (3100 miles).

The spectrum of radiant energy waves we call light, visible to human eye, is very narrow, ranging from approximately 380 nm to 770 nm. Wavelengths shorter or longer than these do not stimulate the receptors in the human eye.

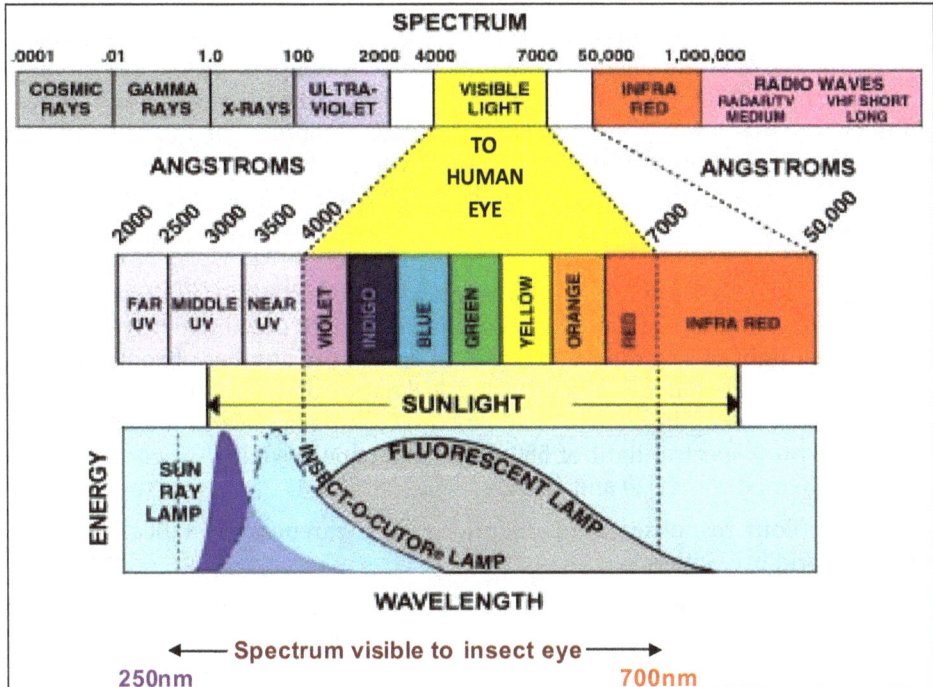

Figure 10: Complete Spectrum and the Part Visible to Insect Eye and Human Eye.

The shortest wave light that serves as a colour stimulus has a wavelength of about 380 nm. The longest wave light that serves as a colour stimulus has a wavelength of about 770 nm. This range (380 nm to 770 nm) is called the visible spectrum (for human eye). Beyond this range is darkness. The term 'colour stimulus' or 'colour' where ever cited in the text refers to the human eye and not to the insect eye. Insect visible spectrum however, extends in UV range below 380 nm upto 250 nm.

Light has numerous attributes three of which are related to the role, light plays as a colour stimulus; these attributes are:

1. Wavelength of single frequency light
2. Intensity
3. Wavelength composition

Wavelength of Single Frequency Light

Wavelength refers to the lengths of the waves in which light travels. Wavelength is usually expressed in mill microns (millionths of a millimeter). The latest term however is "Nanometer" which is recognized by several professional societies as the preferred nomenclature for expressing measurements of wavelength. 1 nanometer (nm) = 1 mill microns (µ) 10 angstroms =10^{-9} meters (m).

Intensity

The intensity of light is related to the rate of incidence of the energy (in units of ergs per second) on to the visual receptor. It refers to the physical energy content of the photo-stimulus applied. To serve as a colour stimulus the intensity of light must be above a certain minimum threshold level. The least amount of energy that evokes a visual response 50 per cent of the time is called the absolute threshold.

Wavelength Composition

A third attribute of light related to its role as a colour stimulus, is its wavelength composition. A colour response may be initiated by light of a single wavelength or frequency of light of several wavelengths in an infinite number of combinations. The wavelength- content describes the physical nature of the stimulus. Two stimuli of different wavelength- content way be equal in appearance to the animal (for example, a pure spectral light of 580 nm. looks equally yellow to us as does an approximate mixture of 520 and 640nm.

Basic colour responses are perceptions (or awareness) specifically of hue, saturation and brightness.

Hue is the dimension of colour referred to a scale of perceptions ranging from red through orange, yellow, green, blue and back to red. The hue of a colour usually depends primarily on the particular wavelength distribution of the light acting on the eye.

Saturation is the dimension of colour representing a colour's degree of departure from an achromatic colour of the same brightness.

Brightness is the dimension of colour representing a colour's similarity to some one of a series of achromatic colours ranging from very dim (Dark) to very bright (dazzling).

Types of Light Sources

The sun (natural) and electric lamps (artificial) are the major light sources transforming energy from another form into the radiant energy wavelengths stimulating our eyes which permit vision, which we call light.

There are four basic types of electrical light sources used today. (i) Incandescent (ii) Fluorescent (iii) high intensity discharge lamps (Mercury Vapour) and (iv) LED (Light Emitting Diodes)- recently introduced - a semi conductor light source.

1. **Incandescent lamps:** Incandescent lamps produce light by electrically heating high resistance tungsten filaments to intense brightness. The common tungsten filament lamp is an everyday example of an incandescent source.

2. **Fluorescent lamps:** Fluorescent lamps produce light by establishing an arc between two electrodes in an atmosphere of very low pressure mercury vapor in a chamber (the glass tube). This discharge produces ultraviolet radiation at wavelengths which excite crystals of phosphor (the

white powder) lining the tube wall. The phosphor fluoresces converting ultraviolet energy at a wavelength (253.7 nm) into visible energy light. Latest edition in fluorescent lamp is of Compact Fluorescent Lamp **(CFL)** using low wattage energy saver, is useful for household purposes.

3. **High intensity discharge lamps:** Light may be emitted when the atoms of molecules of gases or vapors are excited by high temperatures. The emitted spectra are characteristic of the atomic structures of the elements; different gases thus produce different and characteristic spectra. Examples:

   a. **Mercury vapour** in a discharge tube produce a line spectrum and the emitted light appears pale blue to pale green.

   b. **Sodium** produces a line spectra consisting chiefly of two bright yellow appearing lines- produces bright yellow light.

4. **Light Emitting Diodes (LED)** - Details in Chapter 6.

Table 1: The Light Emitting Efficiency of different Types of Lamps

Sl.No.	Type of Lamp	Lumens of Light per watt. (lpw) of Power Consumed
1.	Incandescent general lighting lamps	17-23 lpw
2.	Typical white fluorescent	50-80 lpw
3.	Mercury lamps	50-55 lpw
4.	Multi-Vapour lamps	80-90 lpw
5.	Lucalox (R) Lamps	Over 100 lpw

* Lumen is the unit of luminous flux equal to the flux emitted in a unit solid angle by a uniform point of source of one candle. One lumen per square foot equal 1 foot candle.

5

Visual Response and Spectrum Attractive to Insects

Vision begins with the absorption of photons by the photosensitive substance present in the retinal cells. The retina cannot perform colour vision if all visual cells are of the same type of receptors. The colour vision is possible only if retina contains two or more receptors. The retina will be excited in a different way by radiations of different spectral composition.

Colour vision is the ability to distinguish radiations according to their spectral composition at any value of intensity. Therefore, in studying animal vision it is advisable to use term 'radiation' instead of 'colour' for instance radiation with peak at 585 nm wavelengths instead of "yellow". For up to date surveys on various aspects of insect colour vision reference is made to the works of Mazokhin Porshnyakov (1956, 1964 and 1972), Autrum (1960), Goldsmith (1961), Dethier (1963), Burkhardt (1964) and Nelson (1972).

The question of what spectral region the insect eye is sensitive to is still rather confused, particularly in regard to the long wave end of the spectrum. Comparing the human eye, insects see about the same amount of the red end of the spectrum as we do, but they see much further into the ultraviolet. In precise term, their general range is from 250nm to 700nm, compared with the human range of 380nm to 760nm. Luckiesh (1946) has defined spectral ranges (wavelength bands) by names of spectral colour (human eye) commonly used as follows:

Spectral Colours and their Wavelength Range

Name of Spectral Colour	Wavelength Range (milli micron or nm)
Middle Ultraviolet	200 to 300 nm
Near Ultraviolet	300 to 390 nm
Violet	390 to 430 nm
Blue	430 to 490 nm
Green	490 to 550 nm
Yellow	550 to 590 nm
Orange	590 to 620 nm
Red	620 to 770 nm
Infrared	770 nm to $10 * 10^5$ nm

Responses of insects towards visible and ultraviolet radiation have been studied in two different ways.

1. The effect of the radiation measured by means of some specific behavioral changes in behavioral response.
2. Functional changes in the receptors measured by physical methods (electrophysiological). Behavioural studies have been of more direct interest to economic entomologist.

The radiations having highest attracting effectiveness are the ones which stimulate only the shortwave receptors without exciting the other visual receptors. Such radiations are ultraviolet and violet blue (Mazokhin Porshnyakov, 1963).

Observations on Visual Responses of Insects towards Light

Vision in insect is evolved only once; therefore, insects generally see in three specific colours *i.e.* ultraviolet, blue and green (Birscoe and Chitka, 2001). Based on discussion on visual responses of various insects supported by observation data presented below three distinct attractive regions of wavelength bands have been identified as:

    I. Wavelength region of UV radiation (350 to 380nm)

    II. Wavelength region of visible- violet blue radiation (400 to 480nm)

    III. Wavelength region of visible- greenish yellow radiation (520 to 600nm)

I. Response to UV Radiation (350 to 380nm)

Insects are very sensitive to shortwave radiations, more than any other animal species. The hypothesis that insects prefer ultraviolet light because their eyes are stimulated more strongly by ultraviolet than by any other radiations is incorrect. That stimulates the eye most is not the brightness of the radiation but its colour *i.e.* the spectral composition (Mazokhin Porshnyakov, 1969).

An early observation of insect reaction to radiation beyond the visible was made by Lubbock (1882) "that ants are not sensitive to the ultra-red rays; but on the other hand, that they are very sensitive to the ultraviolet rays which our eyes cannot perceive. Vast amount of scientific literature concerned with visible and ultraviolet radiation and insect responses to it, has established that the range of responses for some insects extends into the ultraviolet region to somewhat below 300 nm. Goldsmith (1961), Response of many insects occurs in the near ultraviolet at about 365 nm (Hollingsworth, 1961, 1964; Weiss, 1941, 1943). Some insect species have shown peak responses to light in the 490 to 520nm range (Goldsmith 1961; Hollingsworth 1961, 1964; Stermer 1959; Weiss 1941, 1943, 1944).

The first authors to investigate the usefulness of incandescent lamps were Parrot (1927) and Peterson and Haussler (1928). They compared the effectiveness of these lamps using different colour filters. They found that the butterflies in particular *Laspeyresia* and other pests clearly preferred violet and blue light. Hamilton and Steiner (1939), Porter (1941) and Gui *et al.* (1942), experimenting with various lights including mercury, confirmed that insects are strongly attracted by the blue violet radiations. Later on, it was ascertained that most effective light source for the collection of insects is ultra-violet radiation emitted by Mercury Vapour lamp (Robinson and Robinson 1950, Taylor and Deay 1950, Frost 1953, 1954 and 1955 and Glick and Hollingsworth 1954 and 1955).

II. Response of Insects towards Visible Violet Blue Radiation (400 to 480nm)

Vaishampayan (1979) investigated in detail the spectral specific responses of *Heliothis armigera* moths in a visible spectrum of 400 to 700 nm wavelengths under controlled conditions of light source in laboratory following the same techniques described earlier by Vaishampayan *et al.* (1975). Observation data are summarized in Table 2 below:

Table 2: Spectral Specific Responses of *Heliothis armigera* Moths in the Visible Spectrum in the Range of 400 to 700 nm Wavelengths

Sl.No.	Light Sources Tested			Mean No. of Moths Responded (11 Replications)
	Colour of Light visible to Human Eye	Optical Filters Used (Kodak)	Spectral Quality (30 per cent) Cut off Wavelength Band	
1.	Dark violet	K-36	400 to 430 nm and 690 to 700 nm	16.36 a
2.	Violet	K-34 A	420 to 460 nm and 690 to 700 nm	7.00 b
3.	Blue green	K-45	460 to 490 nm	6.09 bc
4.	Deep green	K-61	510 to 540 nm	3.27 d
5.	Deep yellow	K-9	500 to 700 nm	4.18 cd
6.	Red	K-25	600 to 700 nm	4.45 cd
			CD (P=05)	1.91

Results indicated highest response of moths towards dark violet light (400 to 430 nm), being highly significant at 1 per cent level compared to all other sources. The next better sources were violet and blue green ranging between 420 to 490 nm.

Tests were conducted during (1992) against black cutworm *Agrotis ipsilon* moths following similar set of experimental techniques, but the responses were however, little different. Highest response was observed in a wavelength band of 460 to 490 nm (violet blue light) instead of 400 to 430 nm as observed in *H. armigera* moths. The response of the moths was significantly highest in this range compared to all other light sources tested in visible range (Veda and Vaishampayan 1993).

In field studies against *H. armigera, S. litura* and *A. ipsilon* Mercury Vapour 125 watt lamp followed by ultra-violet 15 watt lamp proved most effective light sources (Vaishampayan and Verma, 1983). In 40 watt Fluorescent colour lights, blue colour (450 to 480 nm wavelength range) proved most attractive to all these three species compared to green, yellow, white and red colours (See Table 26).

Why Mercury Vapour Lamp is more Effective than Incandescent Lamp?

Mazokhin Porshnyakov (1956a, c) conducted the experiment to answer this question comparing the attracting effectiveness of 200 watt lamps of Mercury Vapour and Incandescent bulb respectively using two different filters. These filters were standardized against the brightest spectral lines of mercury emitting radiation at 365nm (ultra violet) using UV filter and mixtures of 546 nm and 577 nm (greenish yellow spectrum). The number of insects captured by 200 watt Incandescent lamp was taken as criteria. Comparative data of the observations are presented in table 3 below:

Table 3: Number of Insects Captured by 200-w Light Traps of different Spectral Composition, in Armenia*

Insects	Number Caught in June (9 Nights)			Number Caught in July (5 nights)	
	MV Lamp with UV Filter	MV Lamp with Yellow Filter	Incandescent Lamp	MV Lamp with UV Filter	Incandescent Lamp
	A	B	C	A	C
Butterflies					
Sphingidae	23	3	0	13	0
Noctuidae	768	77	55	464	38
Microlepidoptera	1682	500	335	1875	346
Beetles					
Cicindela	2	2	1	29	1
Heteroceridae	491	7	5	1940	94
Melolonthinae	7	7	4	81	28
Pentodon	28	4	0	12	0
Oryctes	9	0	0	2	0
Bugs					
Corixidae	1985	195	49	2860	76
Reduvidae	31	7	6	9	8

Contd...

Table 3–*Contd...*

Insects	Number Caught in June (9 Nights)			Number Caught in July (5 nights)	
	MV Lamp with UV Filter	MV Lamp with Yellow Filter	Incandescent Lamp	MV Lamp with UV Filter	Incandescent Lamp
	A	B	C	A	C
Hymenopterans					
Ophioninae	2	2	4	4	5
Braconidae	629	538	291	2046	1316
Dipterans					
Doliochopodidae	630	46	15	865	13
Tabanidae	18	4	1	22	1
Ephemerides	89	13	11	714	25
Caddis flies	692	106	46	772	62
Chrysopidae	2	1	2	1	9
Earwigs	2	1	0	13	0
Orthopterans	3	1	0	12	0
Total	**7098**	**1512**	**825**	**12764**	**2122**

*After Mazokhin Porshnyakov (1956c).

Results clearly show that insects are attracted to a greater extent by Ultra Violet than by Yellowish green radiation or mixed radiations produced by Incandescent bulbs. Although a lamp with a filter gives out less light than a lamp without a filter it doesn't reduces the response. It shows that insect prefer one source of light to another one on account of its spectral composition (wavelength content) and not for its brightness.

III. Response of Insects towards Greenish Yellow Radiation (520 to 600 nm)

Besides Ultraviolet and violet blue radiations there is a third distinct attracting wave length band in a greenish yellow region (520 to 600 nm in visible spectrum) known to elicit a strong positive response in whiteflies, aphids and leaf hoppers belonging to order Hemiptera. Paddy gall flies (order: Diptera) are also known to be highly attracted by greenish yellow coloured light emitted by 200 watt Incandescent Tungsten filament bulb.

Vaishampayan, Kogan, Waldbauer and Wooley (1975) studied in detail the visual behavior of Green house white fly *Trialeurodes vaporariorum* conducting series of experiments under controlled conditions and known spectrum of light sources tested. Results demonstrated:

1. Strong positive response of white flies towards greenish yellow colour covering wavelength range of 520 to 610 nm. (Primary peak).

2. Moderately positive response towards Ultra Violet light below 400nm (Secondary peak).

3. Light in the blue violet region between 400 to 520 nm wavelengths inhibited the response, indicating negative phototropic response.

4. In a paired comparison the response of GRWF to UV was nearly five times higher compared to blue source of light. However, when compared with yellow (monochromatic yellow) the response towards yellow was much greater (Figure 11).

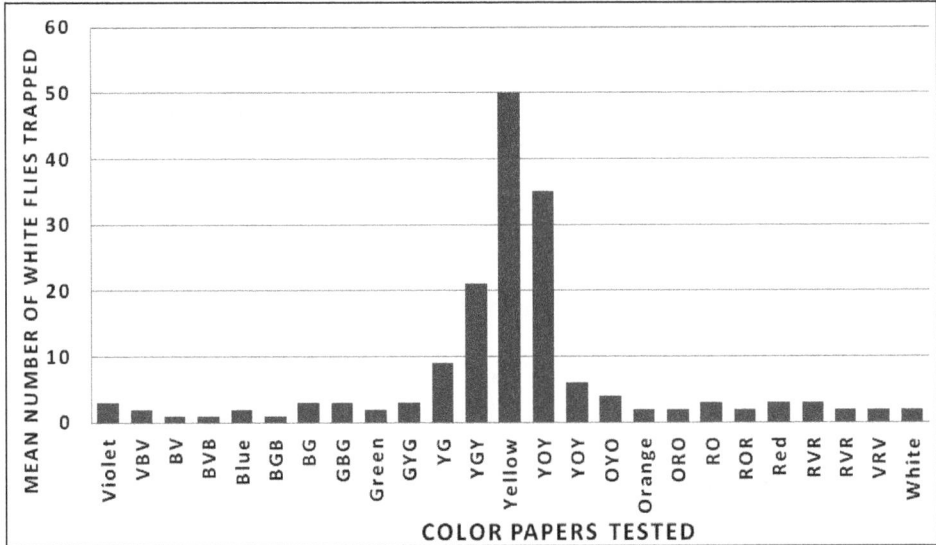

Figure 11: Comparative Response of the Greenhouse Whitefly to Light Reflected by Papers of 24 Hues and White. Test period 1 hour.

Sasamato Kaoru, Kobayashi Masarmi and Shiraishi Hirobumi (1966) investigated the attracting effectiveness of 13 lamps of different wavelengths ranging between 300 nm (UV) and 700 nm, against green leaf hoppers of rice (*Nephotetix cincticeps* Uhler Hemiptera Jassidae).

Black light, BL-lamp (360nm) demonstrated the highest effectiveness in a wavelength region from invisible to visible blue. Blue B-lamp (420nm) indicated less effectiveness (about 42 per cent only) compared to BL lamps. The peak of effectiveness or highest response was however, observed in a visible range at 630 nm (sunlamp).

6
Electric Lamps Used in Light Trap

Electric Lamps and their Characteristics

Numerous lamps have been used singly or in combination to determine photo responses of various economic insects during the past 75 years. These lamps were selected to provide a wide range in radiation output in terms of quality (wavelength) and quantity (power). Most of these lamps were commercially available. Those included were (I) Incandescent 150-w, and 300-w (II) Gaseous discharge 100-w H-4 mercury vapor and (III) 4-w argon glow; and fluorescents 15-w black light (BL), black light-blue (BLB)

There are four basic types of electrical light sources used today- (i) Incandescent (ii) Gaseous discharge (Mercury Vapour) lamps (iii) Fluorescent and (iv) LED lamps.

(i) Incandescent Lamp

The wavelength composition of the energy emitted by an incandescent source depends upon the temperature to which the source is heated *i.e.* the colour temperature.The common tungsten filament lamp is an everyday's example of an incandescent source.

Incandescent lamps produce radiation by heating a tungsten filament. The resulting continuous spectrum includes a small amount of ultraviolet, considerable visible light especially rich in yellow and red, and a peak of radiation in the infrared region which includes about three-fourths of the total lamp output (Figure 12).

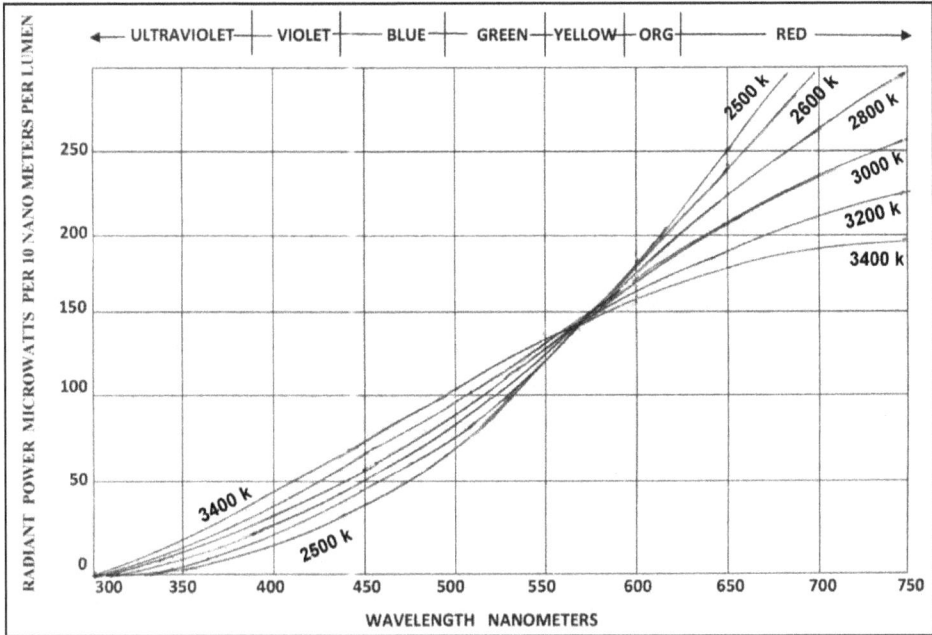

Figure 12: Spectral Energy Distribution of Incandescent (Tungsten filament) Bulb.
Colour temperature varying from 2500º to 3400ºk (Kalvin).
(*Source*: General Electricals, USA).

(ii) Gaseous Discharge Lamps

It produces radiation from excitation of gas molecules in the form of an arc, or plasma, by passage of an electric current through the gas. Each gas produces its own characteristic pattern of colours, giving a non continuous spectrum of bright lines at particular wavelengths. Mercury vapour lamps produce a line spectrum and the emitted light appears pale blue to pale green in colour. It produces primarily ultraviolet, blue, and green radiation with little red (Figure 13).

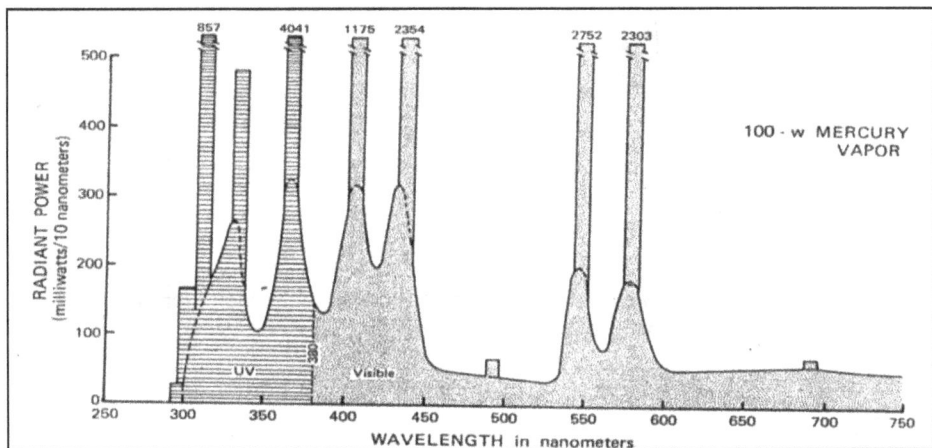

Figure 13: Distribution of Power Output (Spectral Energy) from 100w MV Lamp.

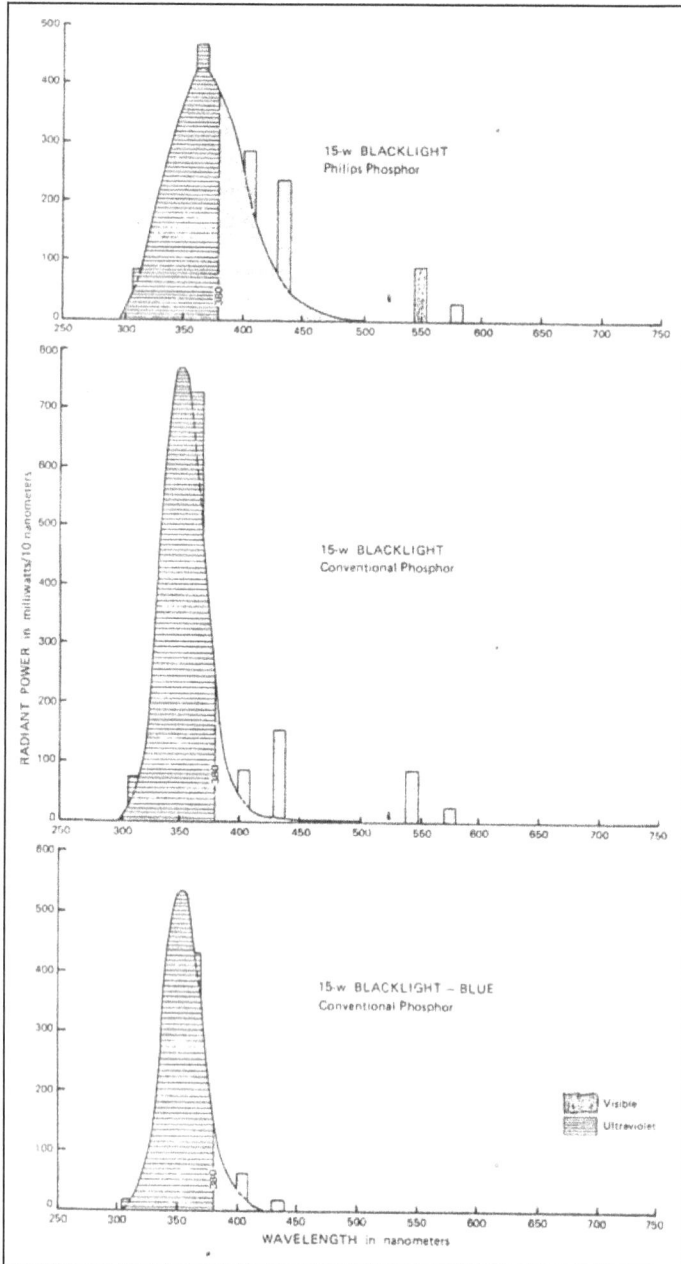

Figure 14: Spectral Energy Distribution of Fluorescent Black Light Lamps BL & BLB with different Phosphor (USDA Bull., 1974).

(iii) Fluorescent Lamps

These are fundamentally a modification of the gaseous discharge lamp. The basic lamp structure is a low-pressure mercury vapor tube, essentially identical to the germicidal lamps tested. The inner surface of fluorescent lamp envelope is coated with various phosphors which absorb short wavelengths and reradiate the energy at longer wavelengths. The light emitted by fluorescent sources generally has a continuous spectrum with superimposed discontinuous spectra from discharges of gases included in the lamps. Certain phosphors fluoresce in the ultraviolet region and others at various wavelengths in the visible spectrum. This includes Black Light (BL and BLB) in UV range, and fluorescent tube lights in blue, green, white (day light) and pink colour in visible range. Black light is a popular name for Ultra Violet lamp radiating energy within the wavelength range of 320 to 380 nm. Fluorescent black light lamps (BL and BLB) use a phosphor which converts the 253.7 nm energy of the basic mercury arc discharge to longer UV wavelengths.

The conventional GE 15-w. BL lamp (F15T8-BL) emits approximately 2.6 watts total radiant power. The BLB lamp varies from the BL lamp only in that it is self-filtered with a red purple bulb. This bulb absorbs the visible light radiated by the BL lamp. The spectral emission curves for three types of fluorescent lamps- 15w Ultra Violet black light BL and BLB are shown in Figure 14.

The 2-w. argon glow lamp is an electric-discharge lamp in which blue, violet, and near ultraviolet radiant energy is generated in the space close to the electrodes. The spectral emission curve for this lamp is shown in Figure 15.

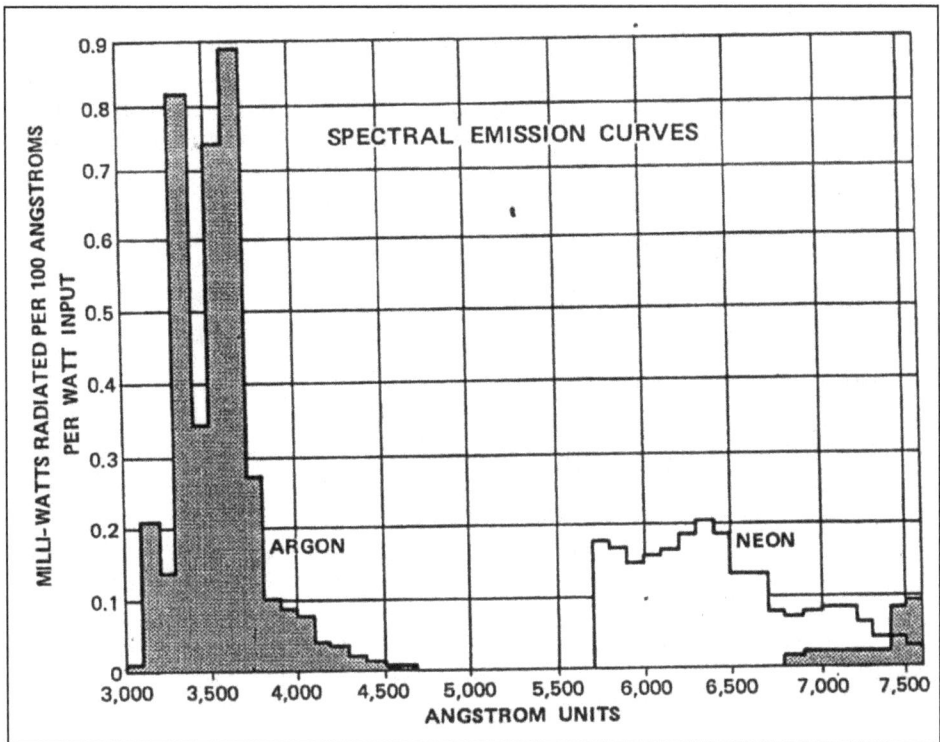

Figure 15: Spectral Emission Curves for Argon Glow and Neon Glow Lamps.

(iv) Light Emitting Diode (LED) Light and its Use in Light Trap

A light emitting diode (LED) is a semi conductor, new type of fourth generation green light source. Originally, LEDs were introduced as a practical electronic component in 1962 and early LEDs emitted low intensity red light.

The LED is a solid state unit that converts electricity to light with minimal generation of heat and therefore is a very efficient source of light. They have some advantages for selective wavelengths, adjustable light intensities, high luminous efficiency, low electric consumption, prolonged life, low weight, compact size and are environmentally friendly (Tamnlaitis 2005, Yeh and Chung, 2009). Very low electicity consumption makes the LED light sources most suitable to operate by the solar energy/battery, including the operation of light trap in remote areas. These advantages make LED light traps a good alternative to the commercial electric lamps presently used for mass trapping and monitoring of phototrophic insects. There is a good scope to use LEDs to control insects in managed environments such as green houses or poly houses or storage godowns (food grains, potatoes and apples storage godowns etc.).

In recent years, manufacturing and technological advances have reduced the retail cost of high efficiency LEDs. Typical modern LEDs produce emissions within a narrow 5-nm wavelength bandwidth. Light Emitting Diode colours ranges from UV (350nm) to infrared (700nm), including colours in visible range *i.e.* violet, blue, green, yellow and red depending on the chemical composition of the LED. Brightness level is determined by the electrical current (measured in milli amperes) passing through the LED. Higher current produces brighter light but the longevity of the bulb is reduced. A LED will function for several thousand hours if not subjected to electrical overhead. The solid state design of the LEDs makes them durable under field conditions; they are difficult to shatter and rarely need replacement.

Potential LED Light Sources for Use in Light Trap

(Presently not available in the market on commercial basis)

    (i)   Ultraviolet- peak at 360nm, 365nm; 4-6W

  (ii)   Visible violet/blue- 410nm, 420nm; 10-12W

 (iii)   Green- 520nm, 550nm; 12-15W

 (iv)   Yellow- 560nm, 580nm and 600nm; 12-15W

Large scale production of LED light sorces in above ranges, specifically the ultraviolet, is urgently needed in India. Such light sources are not commercially available any where in the world for its use in light trap in the farmer's field.

Observations on Response of White Flies towards LEDs in Light Trap

    1. Photo response of white fly *Bemisia tabaci* (Hemiptera: Aleyrodidae) Kim, Yang, Chuing and Lee (2012)

The photo response of the tobacco white fly to light emitting diodes of four different wavelengths and various intensities was tested in an LED equipped Y shape maze chamber and compared with the response to black light (BL), which is typically used in commercial traps. The BL showed the highest attraction rate (90.3 per cent) to *Bemisia tabaci*, followed by a similarly strong attraction to the blue LED (89.0 per cent), the yellow LED (87.7 per cent), the green LED (85.3 per cent) and the red LED (84.3 per cent). These results suggest that energy efficient LEDs could be used for more environmentally friendly insect control.

2. Evaluation of different wavelengths of LED for adults of *Aleurodicus dispersus* Zheng, Zheng, Wu and Fu(2014)

Use of LEDs as an inexpensive light source has been evaluated and examined the relationship between the captured number and the population density of adult *A. dispersus* in the field.

It was found that the violet LED (405 nm) traps significantly higher number of white flies than those of blue (460 nm), green (520 nm), yellow (570 nm) and red (650 nm). The adults captured by light traps equipped with violet LEDs and smeared with liquid paraffin had a significant positive correlation with the population density of adults in a guava orchard, with a correlation coefficient of 0.828. The light traps with 15 violet LED bulbs hung into 550 ml plastic bottles and smeared with liquid paraffin were the portable devices for attraction of adult white flies *A. dispersus*.

A. dispersus exhibits positive phototaxis and light trap is a more appropriate tool for monitoring. Zheng *et al.* (2010) evaluated the response of adult *A. dispersus*. to six LED wavelengths (violet 405 nm to infrared 850 nm) in a hexagon maze and found that the whiteflies preferred violet LED to other five LED wavelengths.

Use of Ultra Violet-Black Light Lamps in Light Traps

Different species of insects exhibits varying degree of attraction to different parts of the Ultra Violet spectrum. Large majority of insects are sensitive to UV-A band between 315 nm and 380 nm. The maximum effectiveness is achieved at about 365 nm. Insects are less sensitive to other parts of the UV spectrum *i.e.* UV-B between 280 nm and 315 nm and least sensitive to UV-C between 100 nm and 280 nm. UV lamps are less selective and may be used to capture a large number of flying insects whether they are dangerous as crop pest or merely passing through the premises. The effectiveness of these traps was demonstrated in studies performed by the CICRP in grain mills in Marselle, France.

Black light (UV lamp) UV-A type available in 15, 10 and 8 watt tubes is a third important light source. The most common black light bulb is the 18" 15-w fluorescent tube (F15T8 BL or F15T8 BLB). 15 watt UV tube is most commonly used light source in USA, radiating energy with peak emission in a spectral range of 360 -365nm. Majority of insect pest species belonging to order Lepidoptera, Coleoptera, Orthoptera, Diptera, Homoptera and Heteroptera etc. are highly to moderately sensitive to UV-A light. Since 8w and 10w UV tubes are available in India in 10" to 12" length size, they are better source than 15w UV (using two tubes at a time with total wattage output at 16 to 20w).

Quantum 368 Black Light Bulb (Sylvania International)

The Quantum 368 black light bulb has been claimed to be more effective than the 350 black light bulbs that are currently used in light traps and night collecting lights. The Quantum black light bulbs are the first major advancement in UVA technology in over 50 years. The Quantum black light bulb was designed around new Phosphor technology that generates extra attraction for pest control in the food service industry. These bulbs are 100 per cent more effective and 40 per cent more powerful than standard 350 black light bulb. Quantum black light bulbs peak at 368 nanometers wavelength. Depreciation of UVA is also significantly reduced resulting in a 40 per cent increase in output over standard 350 black light bulbs. The manufacturer of the Quantum black light bulbs, Sylvania International, states that the Quantum bulbs preserve their output at 70 per cent over 5000 hours. This is due to the re-engineered spectral power distribution of the bulb, giving it a sharpened peak of 368 nanometers, (the optimum level for flying insect eye activity) double than that of any standard 350 black light bulb.

Although designed primarily for the pest control industry, the Quantum Black Light bulbs emits light with an effective attractive travel distance 60 per cent greater than the current 350 black light bulbs. Simply put, the Quantum black light bulbs emit a UV light that extends out farther from the bulb, attracts more moths and has a longer effective bulb life. Quantum Black Light bulbs can be used in place of the standard 350 Black Light Bulbs.

Health and Safety Issues related to UV Lamp

The Black Light traps (BL and BLB) do not present any real danger to persons in proximity since the wavelength of 365 nm is very close to the wavelengths of visible lights detectible by the human eye at about 380 nm.

Regions of UV Light Sources

According to wavelength range Ultraviolet radiation is divided into three distinct bands i.e. UV-A, UV-B and UV-C. Each has different penetration properties and potential for damage.

Band	Also known as	Wavelength Range	Hazard Potential
UV - A	Near UV	380 – 315 nm	Lowest hazard
UV - B	Middle UV	315 – 280 nm	Medium to High
UV - C	Far UV	280 – 100nm	Highest

Harry Katz, (1995) while reviewing some of the myth conceptions regarding the health and safety issues related to Ultra Violet emissions from insect light trap using BL lamps (Myth #7) following reference is quoted by the author (reproduced from the Pest Control Technology (PCT) Magazine, August 1995).

A letter dated April 2, 1993 from the Senior Scientist of Sylvania Lighting Co., Salem, Mass., reports that tests show emissions from black lights in Insect Light Traps are well below Threshold Limit Values (TLVs) of the American Conference of Governmental Hygienists. The TLVs are used nationally and internationally. The black light lamp appears to have no characteristics that indicated likely hazards to human health in either the short term or the long term. The TLV's are provided in units of "mill Joules per square centimeter of surface area (mJ/cm²).

☆ **UV- A,** radiation (380 – 315nm) called as black light (BL lamp), is safe for light trap operation. UV-A is a least hazardous radiation. Above the 315nm wavelength the TLV is always over 1000 mJ/cm² and it steadily climbs above that wavelength indicating that the radiation is less & less hazardous with increasing wavelength (Health Science Division, Columbia University, (Created by Muhammad Akram and Paul Rubock in May 2005).

☆ **UV- B**, radiation (315 – 280nm) called as middle UV, (Tanning) is much hazardous UV radiation. In this wavelength TLV is less than 10 mJ/cm²

☆ **UV- C**, radiation (280 – 100nm), The most hazardous UV radiation (Germicidal). Used in water purifier and pathology labs.

Biologically, the different regions of ultraviolet light have completely different effects. The immediate result of exposing the skin to UV-C and UV-B radiation is erythema or sunburn (tanning effect), while UV-A energy is well above the harmful UVB range of energy (between 320 to 280 nanometers). Certain thresholds of exposure to UV light sources have been established by OSHA* to ensure personnel safety as related below:

☆ 1,50,000 hours of continuous exposure at a distance of twenty feet (20') or 6. 1 meters from a light source;

☆ 1,500 hours of continuous exposure at a distance of six feet (6') or 1.83 meters from a light source;

☆ 40 hours of continuous exposure at a distance of one foot (.3048 meter) from a light source.

Sources: i. General Electric Lighting Company; ii. Philips Lighting; iii. Sylvania GTE Products; New England Journal of Medicine; iv. *Occupational Health and Safety Administration (OHSA).

7
Factors Affecting Light Trap Catches

Success of light trap as IPM tool depends solely upon its trapping efficiency *i.e.* attracting maximum number of adults present in the environment and trap them all (ideally 100 per cent) in a collection bag successfully. Killing of these adults efficiently using a fumigant and preserve them in good condition is equally important for use of light trap in surveillance studies. If trapping is insufficient it would aggravate the problem by inviting more adults in the trap area. Therefore, the knowledge of the factors affecting the size and quality of trap catches is extremely important in pest control operation and in ecological studies as well.

Many workers in the past have investigated the influence of various factors on the trap catches of insect pest specially the Lepidoptera. Influence of weather factors and moon light on the trap catches have been studied earlier by Nemac (1971), Persson (1971), Agee, Webb and Toft (1972), Vaishampayan and Verma (1982) and Verma, Vaishampayan and Rawat (1982). Bowden and Church (1972) and Person (1976) examined the responses of large number of insect species of various orders in relation to the regular changes in light illumination on different days of lunar cycle. Southwood (1978) reviewed in detail the trap design in use and factors affecting the trap catches of insects in general. Hardwick (1968), while describing an efficient trap for noctuid moths, reviewed in brief the principles of light trap design.

The size of the total trap catch is governed by three principle factors namely,

1. The density of adult population (ambient population) present in the trap area.

2. The behavioral phase of adult population (feeding, reproductive or migratory phase)
3. The trapping efficiency (the functional aspects of the trap) influenced by the mechanical design of trap and type of light source used.

Each of the three factors is further influenced by various other sub factors as shown in a schematic diagram in Figure 16.

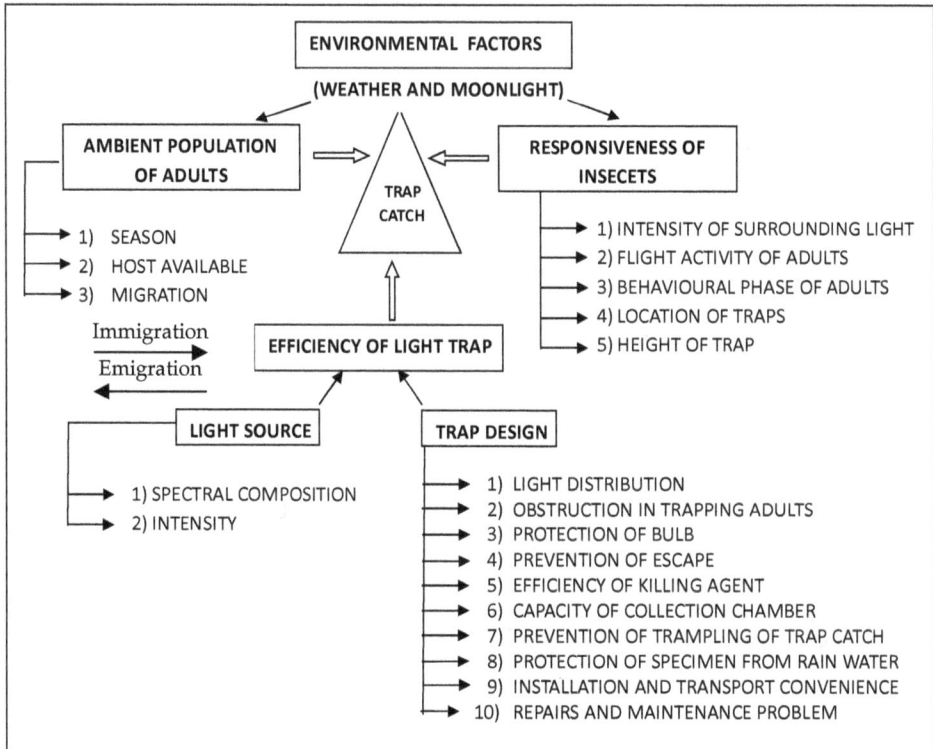

Figure 16: Schematic Diagram Showing Factors Affecting Light Trap Catches of Insects (Vaishampayan, 1985).

The 'ambient population of adults' is influenced by three factors *i.e.* distribution of host plants in the eco system, emergence of adult population in the environment and the migration behavior of the population.

The 'responsiveness of adults' towards light, is influenced by (i) the contrast effect of surrounding light (ii) bad weather comprising of heavy rains, storms, extreme low temperature etc. affecting the flight activity of adults (iii) behavioral phase of flying adults (host finding, mating, ovipositing, or migrating etc. and (iv) the location of trap whether 'in open', or in a place surrounded by buildings or big tress etc.

The **net efficiency of the trap** is the third major factor, a functional aspect which is the ratio of number of adults caught and the total population present in the surrounding area of trap in a unit time. This 'net efficiency' is a product of two principle factors *i.e.* the design of light trap and the type of light source used.

Efficiency of Light Source Used

The type of light source used determines the ratio of insects of a particular species or a group of species attracted towards light trap. Incandescent tungsten bulb emits energy in long wave radiation (greenish yellow, yellow and orange) which is most attractive to leaf hoppers, whitefly, aphids, gall midges etc. but least attractive to Lepidoptera and Coleoptera. The mercury vapour lamps emit energy in beams characterized by both shortwave (UV) and long wave radiation and hence most useful for variety of insect pest species. For more review see chapter 6. Compared to incandescent and UV lamps, MV found to be highly effective light source for very wide range of harmful insect pest. In a comparative study Mercury Vapour bulb, with distinct peaks in UV and visible radiation was found to be the best light source in attracting adults of *H. armigera, A. ipsilon* and *Spodoptera litura* (Vaishampayan and Verma 1983, see Table 26). The consumption of electricity in this lamp is however very high (80 to 125 watts) compared to only 15 watt in UV lamp.

For more details on Black Light UV lamp trap see Chapter 6.

Design of Light Trap Used

The design aspect of trap is of utmost importance in the scientific investigations related to the use of light trap in ecology or pest control studies where quantity of trap catch as well as the quality is of significance. The objective in designing the trap is to increase the total quantum of catch to an optimum level and maintain it in a good condition, without mutilation and safe from rain water, so that the insects could be identified easily and counted correctly. Following points should be taken into consideration while designing a light trap.

1. Light source should be as open as possible radiating light in all directions freely. Rain shed reduces light distribution considerably.
2. The obstacles should be minimum between light source, baffles and the way to collection chamber.
3. Protection of light source (bulb) from rains, storms and theft should be full proof.
4. Escape (returned flight) of insects from collection chamber should be minimum.
5. Sufficient space in collection chamber should be provided to accommodate large collection.
6. Anaesthetizing or killing of insects should be quick. Dichlorivos (DDVP) as fumigant, available in 100 ml bottle is an excellent killing agent.
7. Rain drain system should be so effective that it drains out the water easily.
8. The maintenance and repairs should be minimum and easy.
9. Cost should be moderate.

10. Size and shape of trap should be concise, easy to transport and install at any place.

11. Trap operation should be shock proof.

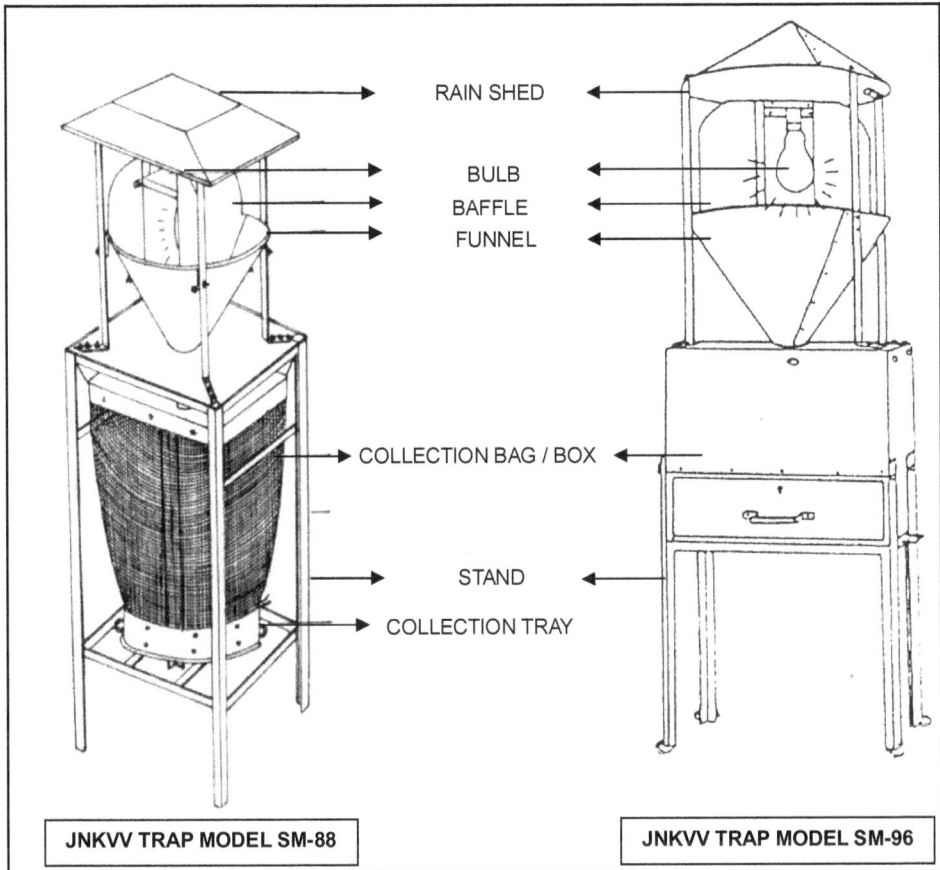

RAIN SHED

BULB
BAFFLE
FUNNEL

COLLECTION BAG / BOX

STAND

COLLECTION TRAY

JNKVV TRAP MODEL SM-88

JNKVV TRAP MODEL SM-96

Figure 17: Designs of Light Trap Developed by Vaishampayan at Jabalpur.

Designs of light traps, considering all the points discussed above have been developed at Jabalpur widely used in field studies for several years (Figure 17). Latest designs developed by Vaishampayan in 2014 suitable for production in plastic mould for farmers use in field are described in Chapter 9.

Effect of Environmental Factors

1. Influence of Weather Factors on the Light Trap Catch

Verma, Vaishampayan and Rawat (1982) studied the effect of three weather factors namely temperature, relative humidity and rainfall by analyzing the trap catches of *Heliothis armigera* moths during the years 1977, 1978 and 1979. The traps were operated daily in active season from January to May. Since the brightness of moon light around full moon is known to suppress the trap catch significantly, due corrections were made to nullify this effect by discarding the data of trap catches collected on two days before and 2 days after the full moon period respectively.

Overall results proved that none of the weather factors studied, had any significant effect on the flight activity of the pest. Hence, the size of the trap catch in relation to its ambient population remained unaffected by these weather factors. Rains, after a dry spell of long period, had a significant but short lived effect showing sudden increase in the trap catches. Results further suggests no need of correction in the weekly moving means of trap catches due to the weather effects in the context of comparing them with the actual population of the pest present in the environment.

2. Influence of Moon Light on the Light Trap Catches

Many of the workers in the past have reported a significant reduction in trap catches of noctuid moths on full moon and around compared to the no moon period (Nemac, 1971; Agee, Webb and Toft, 1972) Most of them compared the responses on full moon and no moon and their conclusions were based on the data of one season only. Bowden and Church (1973) and Persson (1976) made extensive studies examining responses of few insect pest species in relation to the regular changes in light illumination on different days of a lunar cycle. Vaishampayan and Shrivastava (1978) analyzed the response of *Spodoptera litura*, covering 12 lunar cycles of a year and reported significant inverse correlation ship between the trap catch and degree of moon phase. The response of the moths was found to be sharply declined on Full moon day as compared to the No moon day (Figure 18).

Vaishampayan and Verma (1982) carried out detailed investigations on this aspect, analyzing data of trap catches of *Heliothis armigera* adults consistently for four years during the period 1975-76 to 1978-79. The data of trap catches covering only the active period (January to May) of the pest species were analyzed. The intensity or brightness of moon light for each lunar day was measured in terms of degree of moon phase or the relative area of the moon disc which is illuminated. The 'Full Moon' was considered of 360° phase and the no moon as 0° phase. With the division of 360 by 15, each day represented change of 24°, increasing in ascending phase and decreasing in descending phase of a lunar cycle. Abstract data of 4 years observation (mean of 19 lunar cycles) are presented in Table 4.

Differential Response of Insect in Relation to Scotophase

Further observations revealed a consistent difference in the response of moths in ascending and descending halves of the lunar cycle. The data abstracted in Table 14 indicated a higher response of moths in descending half compared to the ascending one in all the years. This differential response is explained on the basis of the association of scotophase (absence of moon on horizon). Since the moths are most active in the early part, the presence or absence of moonlit period in early hours makes a difference in the response. In descending half of lunar cycle, the scotophase is associated with early hours of the night. The absence of moonlit period in early

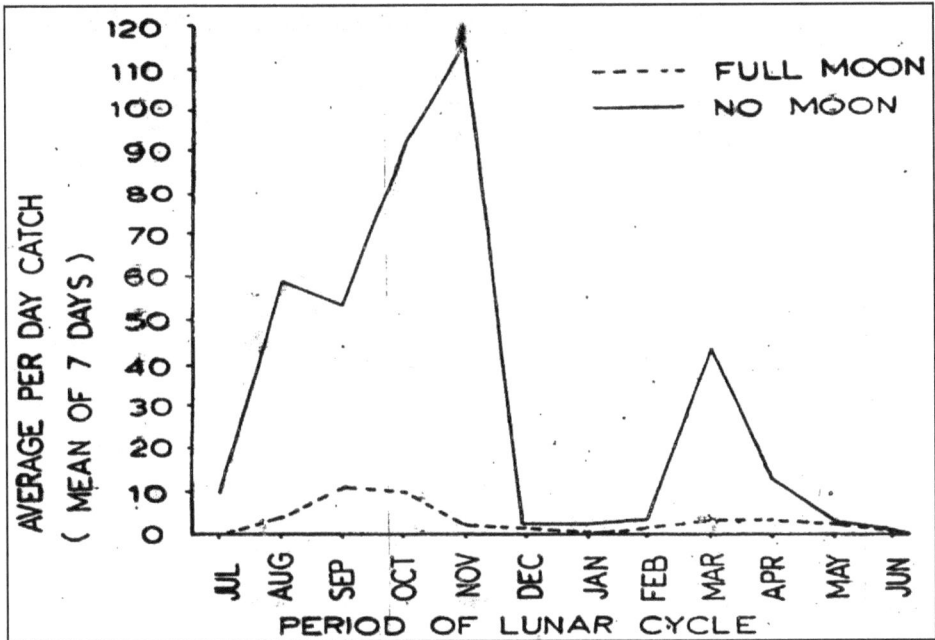

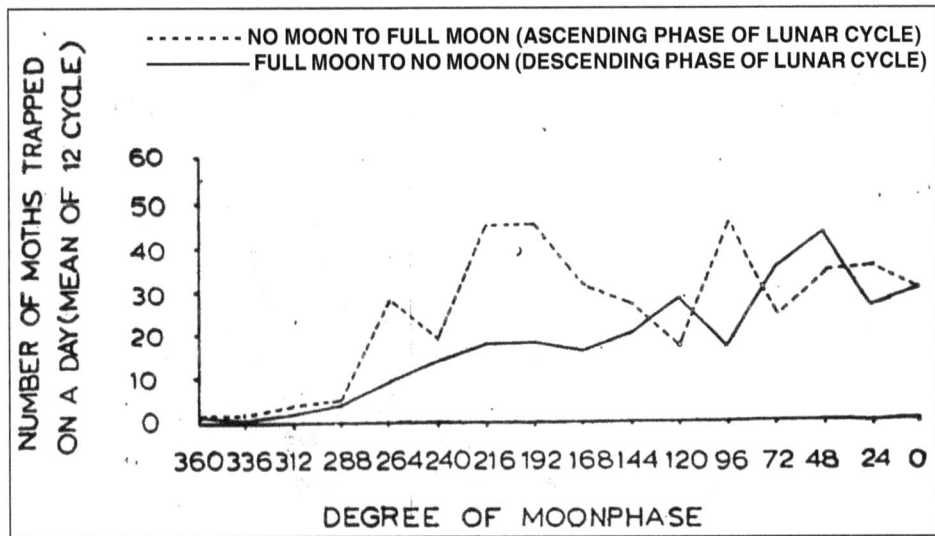

Figure 18: Effect of Moon Phase on the Light Trap Catch of
Spodoptera litura at Jabalpur (1974-75).

hours therefore increases the trap catches, on the contrary, in ascending half it is associated with the late hours after midnight (Figure 19).

Table 4: Relative Response of *Heliothis armigera* Moths in Light Trap Catches on Various Lunar Days in a Lunar Cycle. (Mean of 4 years data 1975-76 to 1978-79) (Mean of 19 lunar cycles)

Ascending Phase (No moon to Full moon)			Descending Phase (Full moon to No moon)		
Lunar Day	Degree of Moon Phase	Average Catch/ Night/Lunar Day	Lunar Day	Degree of Moon Phase	Average Catch/ Night/Lunar Day
	X	Y		X	Y
1	24	71.8	16	336	5.2
2	48	57.8	17	312	5.0
3	72	52.1	18	288	5.7
4	96	40.5	19	264	7.0
5	120	38.7	20	240	29.6
6	144	32.4	21	216	37.0
7	168	27.6	22	192	30.7
8	192	18.7	23	168	38.3
9	216	22.7	24	144	47.0
10	240	21.5	25	120	56.2
11	264	23.7	26	96	63.2
12	288	23.6	27	72	58.8
13	312	16.3	28	48	38.9
14	336	12.6	29	24	47.3
15	360	10.2	30	0	79.6
Corrl. "r"		−0920**			−0.920**

The results show a significant difference in the response of moths between moonlit nights and dark nights. In all the lunar cycles the response was consistently very low on bright moonlit nights around full moon representing only a fraction of the natural population.

Table 5: Effect of Moon Phase Cycle on the Trap Catches

Observation Year with Lunar Cycles Covered	Overall mean per night catch (Mean of 15 lunar days/lunar cycle)	
	Ascending Phase (No moon to full moon)	Descending Phase (Full moon to no moon)
1975-76 (6 Lunar cycles)	24.00	27.45
1976-77 (4 lunar cycles)	5.40	9.66
1977-78 (5 lunar cycles)	88.06	96.00
1978-79 (4 lunar cycles)	8.09	9.41
Mean of 4 years (19 Lunar cycles)	31.34	36.63

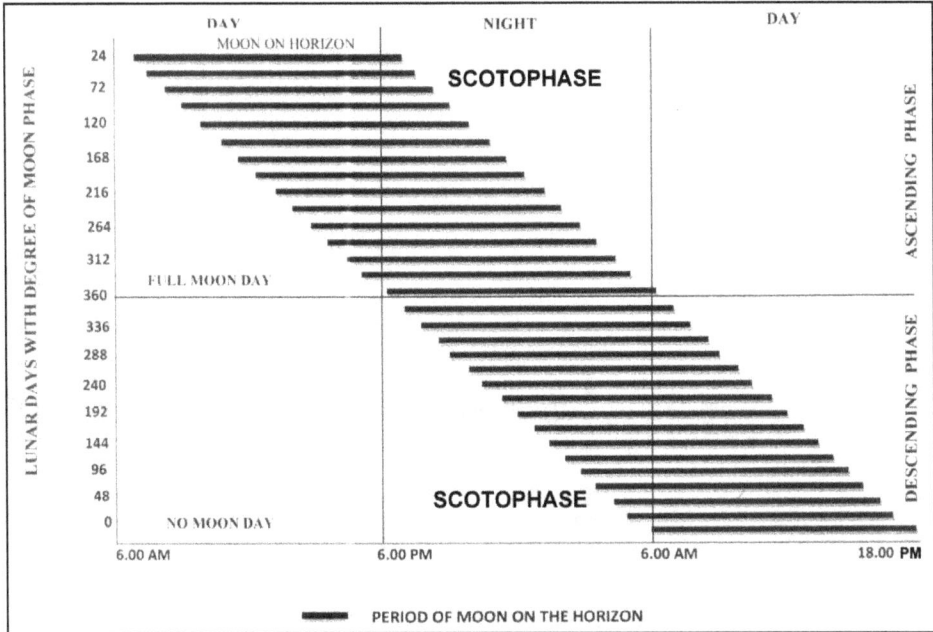

Figure 19: Timings of Moon on Horizon Shown by Horizontal Bars on different Lunar Days of a Lunar Cycle.

8

Principles and Practice of Light Trap Operation

Light trap works on a principle of attraction of night flying adult insects towards artificial light source and trap them in a device called light trap. Philosophy in use light trap is 'to control the pest population in adult stage in reproductive phase prior to origin and spread of infestation in field, at time and place when they (adults) are most concentrated immobile and accessible (CIA factors)'. Light trap concentrates adult population present in the environment arrest their flying movements and make them immobile by trapping in a collection box and kill them using a fumigant (Dichlorvos). The potential infestation of a pest species is concentrated in the ovarioles of female moths in oocyte form (egg stage). Mass trapping of such adults reduces the infestation in field significantly. Thus it is totally an adult oriented approach of IPM.

Adult Oriented Pest Management Strategy

It is a new concept of pest management proposed and discussed by Vaishampayan (2002) which emphasizes to concentrate or orient all control measures towards the adult stage alone instead of targeting the immature larval stages in a crop ecosystem. Adult stage decides the fate of its population in the area governed by reproductive behavior and flight movements of the adults. Theme revolves around suppression of pest population through mass trapping and killing of adults using their behavioral responses *i.e.* visual, olfactory, gustatory, sexual, reproductive and biological etc.

Adult is a most vital stage in a biotic cycle of an insect species which is an end result of a current generation passing through egg, larval and pupal stages encountering over 95 per cent (max.98 per cent) natural mortality between eggs to adult stage. The population explosion is in fact, hidden in adult stage. Pradhan (1973) has precisely pointed out the importance of survival of adult stage in population ecology (Figure 20).

In a static population, the hypothetical rate of survival in a generation, from egg to adult stage, is estimated around 2 per cent only (at least one female + one male). In other words, each female is replaced by one female adult potential to carry the generation with the same egg laying capacity.

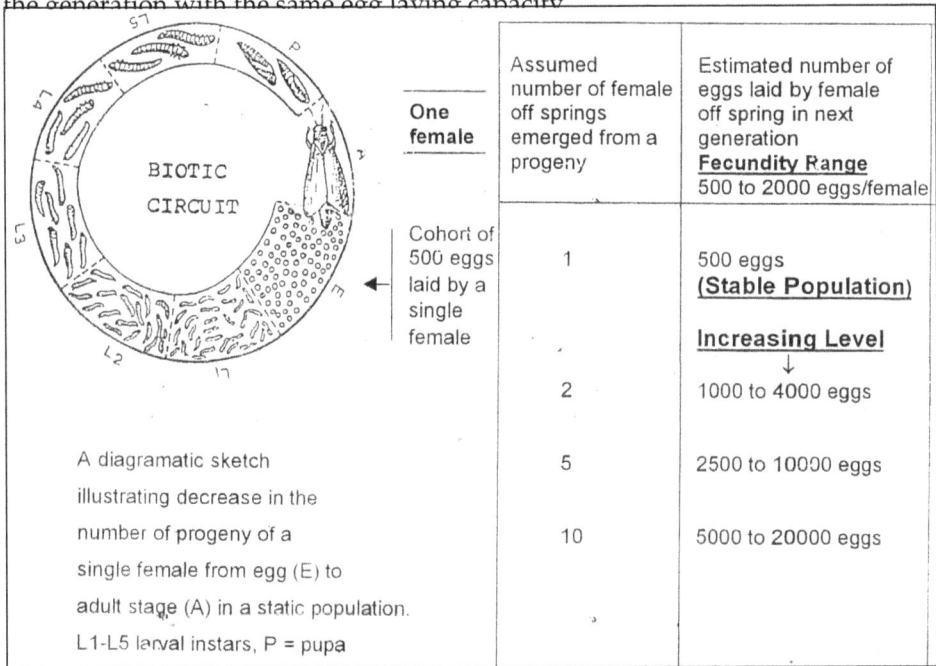

	Assumed number of female off springs emerged from a progeny	Estimated number of eggs laid by female off spring in next generation **Fecundity Range** 500 to 2000 eggs/female
One female		
Cohort of 500 eggs laid by a single female	1	500 eggs **(Stable Population)** **Increasing Level** ↓
	2	1000 to 4000 eggs
	5	2500 to 10000 eggs
	10	5000 to 20000 eggs

BIOTIC CIRCUIT

A diagramatic sketch illustrating decrease in the number of progeny of a single female from egg (E) to adult stage (A) in a static population. L1-L5 larval instars, P = pupa

Figure 20: Importance of Female Adult in Population Dynamics of a Pest Species (Pradhan, 1973).

Addition of every single female over this one unit will increase the infestation level in next generation in a multiple of number of female adults added. Killing of 10 female adults means controlling nearly 10,000 early instar larvae likely to be emerged after oviposition and hatching of estimated 10,000 eggs laid by 10 females, supposed to be spread out in an acre of crop area. Since female need multiple mating prior and during oviposition period, the population of male adults also play a vital role in determining the size of future generation.

The Uses of Light Traps in Various Fields

Light traps serve as an important and valuable method in collecting crepuscular and nocturnal insects for taxonomic purposes, for detection of the presence of insect pest, to determine population changes and help in prediction. It serves as a valuable tool in conducting studies/research on variety of basic as well as applied aspects of practical use in various fields as described in brief below:

A. General Basic Studies

a. Collections for Taxonomic Studies

Useful for faunastic survey and collection of insect pest of various groups including moths, butterflies, beetles, hymenopterans, wasps and dipterous flies, odonates and various parasitic and predacious insects of bio-control importance. It provides knowledge of existing fauna of most important species of crop pests present in the region. Such collections are maintained in museums at national level as well as in the colleges for practical use of the graduate students. University of Illinois, Urbana, U.S.A. offers a course on Insect Taxonomy at M.Sc. and Ph.D. level in summer semester. As requirement of course work students travel for 10 days- Illinois to New Mexico. They operate the light traps in field every night and make collection of insects of various orders and preserve the specimens in collection boxes. Senior author attended this course in summer of 1971.

Note on Change in Generic Name Heliothis to Helicoverpa using Light Trap Collection

For many years, the gram pod borer or cotton bollworm, a major pest of rabi pulses and cotton throughout India has been referred to as *Heliothis armigera* Hubner. Recent investigations by the taxonomist of *Heliothis* fauna questioned the validity of generic name of Heliothis. Hardwick (1965 and 1970) redefined the genus *Heliothis* and described the new genus *Helicoverpa*. The species of *Helicoverpa* were separated from *Heliothis dipsacea* (Linn.).

To determine the species complex of gram pod borer 11 specimens of moths each differing in size and colour pattern of fore and hind wings were sent to Dr. Hardwick at Canada for identification. Specimens were selected from three years collection of light trap catches at J.N.K.V.V. Jabalpur. Ten out of 11 specimens were identified as *Helicoverpa armigera* (Hubner) and one as *Heliothis peltigera*. This is the first record in India for change in generic name from Heliothis to Helicoverpa (Vaishampayan,1977).

Structurally the genus is readily distinguished from *Heliothis* by (i) the possession of a multi coiled vesica in the male and (ii) an alternately dilated and constricted appendix *bursae* in the female (Vaishampayan, 2000).

b. Collection of Live Adults for Biocontrol and Ecological Studies

i. For mass culture of host larvae (Heliothis/Spodoptera) spp. for production of viruses NPV and Bacillus culture.

ii. For observations on reproductive status and matting frequency etc. with reference to studies on migratory behavior and other ecological studies.

B. Studies on Applied Aspects: Use of Light Trap as IPM Tool

i. Detection of insect pest present in the area.

ii. Study of seasonal activity of pest species.

iii. Monitor peak of activity period for predicting potential infestation and need for control measures.

iv. **Direct control of pest population** of major insect pest species

    a. In agricultural crops, vegetable and fruit crops, spices and plantation crops, forests trees like sal, teak, eucalyptus and salai.

    b. Crops grown in green house and poly house.

    c. In storage godowns (food grains, apples, potatoes etc.).

    d. In organic farming including cultivation of medicinal plants where use of pesticides is prohibited.

    e. In cattle yards, dairies to control blood sucking *Tabanid* flies, house flies and mosquitoes.

C. Use of Light Trap in Fisheries

Various terrestrial and aquatic insects collected in light trap serves as artificial feed to the fishes cultured in the ponds (details in Chapter 11).

The Principles of Light Trap Operation

1. Operation of light trap is principally based on the attraction of night flying insects towards artificial light source. Naturally, its operation shall be more successful against the species where adults are positively phototropic in nature.

2. The success of light trap in pest management depends upon how efficiently the adult population is withdrawn from the environment. Ideally the winged females must be caught before they lay eggs on the host plants and the males must be caught before they have mated.

3. Trapping mechanism is an important aspect of trap design. If trapping is insufficient it would aggravate the problem by inviting more adults (untrapped) to the area (details in Chapter 7). Quantity of trap catch *i.e.,* the number of adult collected in the trap is an important factor deciding success of pest control operation. Quality of trap catch is of secondary importance.

4. There is a certain time lag between adult emergence and oviposition (Pre-oviposition period). More this period, greater are the chances of trapping adults before they lay eggs. In general, time lag between adult emergence and newly hatched young ones is short enough and hence the prediction of buildup of pest population in field based on adult activity shall be more reliable.

5. Principally use of light trap shall be most successful against the pest species characterized by following features.

    i. Species have annual life cycle, passing through egg, larval, pupal and adult stages and completes only one or two generations in a year. Adult emergence is limited to a short duration around the period of pre monsoon and early showers. In spring season or beginning of summer season first generation adults emergence from diapausing or hibernating pupal stages starts early in the active season and is completed within a short period of one or two weeks as shown in Figure 21 illustrating seasonal history of white grubs completing only one generation in a year. Excellent results can be achieved if trapping operation is synchronized with this period coinciding with a peak of adult emergence during month of June and July. This leads to collection of newly emerged adults in very large number before

Figure 21: Seasonal History of White Grub-Completing Only One Generation in a Year (June to May) (Yadav and Vijayvargiya, 2000).

they start egg laying in field in crops or in orchards etc. Examples: red hairy caterpillar, codling moths, wood borers and white grubs.

ii. Breeding of pest and its infestation is confined or restricted to certain areas or pockets known to be endemic areas and larval damage is restricted in germinating or early growing stage of crop. Adults arrive in t he area early in the season (as migrants) before the start of normal sowing period in winters. Example: Black cutworm *Agrotis ipsilon*.

iii. Females have profuse egg laying capacity in major active period. Such moths being heavy bodied are easily caught in the traps.

iv. Females have longer pre-oviposition period and hence chances of trapping them are more before they lay eggs and spread the infestation in field.

v. Principally light trap works selectively against nocturnal insects highly phototropic in nature, while most of the beneficial natural enemies including Hymenopteran and Dipteran parasitoid, being diurnal in nature, are spared from trapping. This is the most important behavioral aspect of light trap operation which qualifies to designate light trap as an eco-friendly pest control device.

Practice of Light Trap in Pest Management

Control of insect pest populations using light trap is an age old practice. In early days, before the invention of insecticidal properties of DDT, BHC etc. in 1940, light trap were very successfully used in pest control studies. Almost a century ago, outstanding work was done in India against hairy caterpillars and cutworms. With the improvement in trap designs and sampling techniques light traps were successfully used in studying seasonal activity and population dynamics of many insect pest species all over the world.

Williams (1939) has summarized three different aspects of the uses of light traps. They are as follows: Firstly, for collection of specimens for museum and hence the study of the fauna of special areas, the geographical distribution of insects, and as a source of material for the study of taxonomy. Secondly, as a sampling method to obtain an estimate of the numbers of any species active at any particular locality or date; as an aid to the study of the biology of insects; and in particular to their dates of appearance, relative length and size of broods, migration, the relation of their abundance to the conditions of the environment and the study of the changes in population from place to place, from year to year, in fact what is now called Ecology. **Thirdly**, traps have been used as a method for wholesale destruction of pests with a view to the reduction of damage caused by them.

Monitoring Seasonal Activity and Prediction of Infestation Level

Sampling insects from the environment is very important in entomological studies for indexing their natural population. Light trap serves as an important survey tool to sample population of night flying adults very effectively. Monitoring seasonal activity of adults particularly of ovipositing female, provide very useful information for predicting infestation level of pest early in early instar stages.

With the revival of interest in nonchemical methods of control and more emphasis on ecological consideration, use of light traps has gained wide spread importance in integrated pest management strategies all over the world. Several workers in the past have studied the seasonal activity and relative abundance of adults of many species of crop pests, using light trap and correlated the activity with field infestation in few species (Otman, 1964; Kovitvadhi and Cantele, 1966; Ishikura, 1967; Atwal, 1969; Beckman, 1970; Vaishampayan, 1980; Vaishampayan and Bahadur, 1980; Vaishampayan *et al*, 1981). Its usefulness in studying population dynamics of boll worms was emphasized by Hartstack *et al*. (1973). Blair and Catling (1974) and Betts (1976) reported extensive use of light trap in forecasting out breaks of African army worm *Spodoptera exempta* in Africa. Ishikura (1967) in Japan used traps in the assessment of field population of rice stem borer moths. Yoshimiki (1967) studied seasonal abundance of rice stem borer moths using standardized light trap in about 6340 hectares of rice crop in Japan.

The light-trap catch may indicate the presence of an insect in a given area in potentially damaging numbers. As mentioned by Pfrimmer (1961) light-trap data are used in timing the beginning and end of insecticidal applications for the European corn borer and the corn earworm. In Wisconsin recommendations for the control of corn earworms in canning sweet corn are timed by light-trap catches. Light-trap collections have been used for the past 5 years to predict when insecticidal applications are needed for controlling the tomato fruit worm in Indiana. Timing of the first treatments for codling moth control in Wisconsin in 1959 was determined from light-trap catches. Morris has used light-trap catches of carpenter-worm moths on which to base the timing of trunk spray applications for the control of carpenter-worms in hardwoods in Mississippi.

A study of the published data indicates the possibility of associating light trap data with field infestations for predicting the need of control measures for the following insects: European corn borer, corn earworm, cotton bollworm, tomato fruit worm, mosquitoes, sand flies, codling moths, bud moth and leaf roller in apples, Asiatic garden beetle, white grubs, pink bollworm, armyworms, cutworms, cabbage loopers, and leaf- hoppers. The greatest need is for more investigations on the relationships between light trap catches and field populations and the factors affecting this relationship. Research is needed to determine the size of a light-trap catch necessary to indicate an economically damaging or potentially damaging infestation present in the area.

Texas Agricultural Extension Services has used light trap catches successfully throughout Texas to forecast timing of future generations of *Heliothis* in cotton (Hartstack *et al*.1977). As a public service the Plant Protection Programs, Animal and Plant Health Inspection Service, U.S. Department of Agriculture, issues a weekly Cooperative Economic Insect Report (CEIR). Electric light trap catches of economically important insects were published in a few instances as early as 1952 in the CEIR. A special section on light trap collections of nine species of insects was initiated in 1955 and has continued since that time. In 1963, the number of insect species included in the report was increased from nine to 20, all Lepidoptera. Detection of other insects collected in light traps is also listed in other sections of the report. Insect traps utilizing BL lamps rapidly became valuable tools to entomologists and others in determining the time of appearance and the seasonal abundance of important insect pest species (Pfrimmer (1955 and 1957), Stanley and Dominick (1958), Oatman (1957), Tashiro and Tuttle (1959), and Smith (1962)). However, several different kinds of BL traps were then being employed in general insect pest surveys in the United States.

Use of light trap in activity study of adults is based on the assumption that changes in trap catches are mainly an expression of variation in the actual population of adults present in the environment (Persson,1976). This assumption is however not correct in Toto. It holds well only if due corrections are made in trap catches for effect of certain factors. Besides the influence of bright moon light the degree of correlation between trap catches and field infestation level is influenced by sex ratio and reproductive condition of females and the time lag between adult emergence and appearance of early instar larval stages, specifically in rabi season in central and north India where winter temperatures are very variable. Due consideration, therefore, must be given to these factors to interpret the data more precisely and correctly.

After years of observations at JNKVV Jabalpur, an improved design of light trap developed for efficient collection of trap catches and development of a technique to evolve methods of using correction factors were developed which made it possible to correlate the data of trap catches and field infestation with good precision. Better precision (with higher value of 'r') was obtained when trap catches were subjected to correction factors for (i) moon light effect (ii) sex ratio and (iii) reproductive condition (ovary development of the trapped females).

Data on estimation of correction factors for the effect of moon light on trap catches of *Heliothis armigera* (as a sample case) are presented in Table 6. Data on use of these corrections along with other correction factors in correlating the trap catches with the field infestation of *H. armigera* are presented in Table 7.

A profile of weekly moving means of trap catches provides fairly good estimate of relative abundance of pest species in a season. The short-term effects of weather factors are averaged out in weekly moving means. Monitoring appearance of peaks in light rap catches during principle active season of the pest helps to detect the potential danger of population explosion well in advance, and decide the proper time of application of insecticides synchronizing with these peaks. The objective is to achieve the satisfactory control of pest population with minimum use of pesticides and least disturbance to the ecosystem.

Three years data of weekly distribution of light trap catches of *Heliothis armigera* collected in three different years of infestation levels *i.e.*, epidemic, moderate and low infestation respectively, presented in Figure 22 reflects the seasonal activity of the pest in field very clearly.

Peaks in trap catches with variation in size of peaks as per the infestation level observed, provides very useful information for predicting the potential danger of pest attack and take the necessary control measures to protect the crop well in time.

Table 6: Correction Factor for Influence of Moonlight on Light Trap Catches of Heliothis armigera at Jabalpur (23"N). Vaishampayan and Verma (1982)

Lunar Group No.	Lunar Days Group-wise	Mean Trap Catch/ Lunar Group (Mean of 19 lunar cycles)	Relative Response (per cent) Compared to No. Moon* Days (RR)	Correction Factor* 100/RR
1.	28th, 29th,30th 1st,2nd,3rd,4th (around "No Moon" period)	55.42	Absolute response 100.00	1.000
2.	5th to 7th	32.90	Suppressed response 59.36	1.684
3.	8th to 10th	20.90	37.71	2.651
4.	11th to 13th	21.20	38.25	2.614
5.	14th, 15th, 16th, 17th, 18th (around "Full Moon" period)	7.74	13.96	7.163
6.	19th to 21st	24.70	44.56	2.244
7.	22nd to 24th	38.60	69.64	1.435
8.	25th to 27th	59.40	Absolute response 100.00	1.000

* Based on estimated reduction in trap catches due to moon-light effect.

Table 7: Relationship between Light Trap Catch and Larval Population of *Heliothis armigera* (1985-86)-data for Various Influences

Week no.	Total Catch/ Week/Trap	Catch Corrected to Moon Phase Effect	Estimated Females (Based on sex ratio)	Reproductive Index (R.I.)	Estimated Fertile Females N x R.I.	Early Instar Fertile/5 m² (End of week)	Weekly Mean Min. Temp. °C
	X_1	X_2	N	—	X_3	Y	—
Dec 52	16.0	118	56	1.00	56	116	10.7
Jan 1	1.4	7	4	0.25	1.0	99	2.5
Jan 2	2.0	2	1	0.50	0.5	67	5.8
Jan 3	5.0	15	8	0.72	6.0	17	11.1
Jan 4	5.4	42	23	1.00	23.0	38	12.5
Feb 5	50.0	127	63	0.73	46.0	43	11.6
Feb 6	83.0	83	40	0.50	20.0	27	14.1
Feb 7	29.0	70	35	0.43	15.0	22	11.6
Feb 8	22.0	256	78	0.48	37.0	32	12.9

Comparison of degree of correlation:

(i) Comparison	X_1 v/s Y	X_2 v/s Y	X_3 v/s Y
(ii) Correlation coefficient 'r' value	0.174	0.40	0.81

Remark: 'r' value increase from 0.174 to 0.81

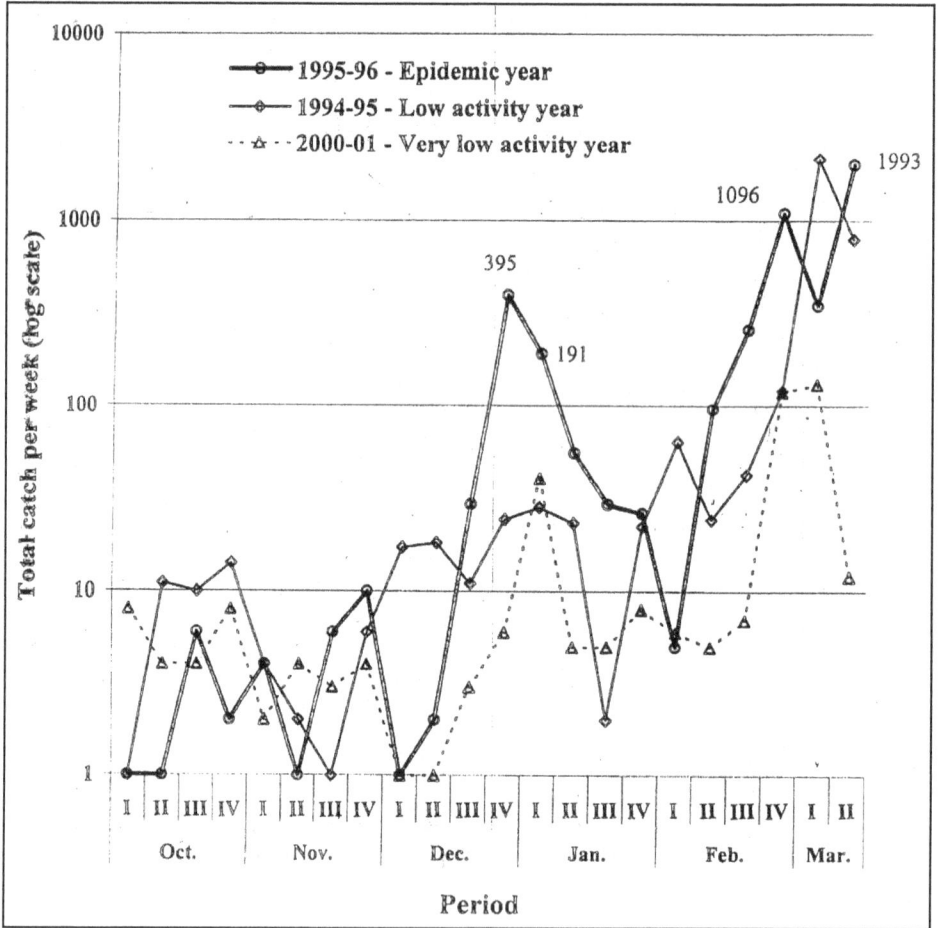

Figure 22: Weekly Distribution of Light Trap Catches of *H. armigera* in different Years of Infestation Levels (High, medium and low activity).

Weekly and Monthly distribution of trap catches provides a good idea of seasonal changes in the activity of the pest. As a sample case, observations made by Verma and Vaishampayan (1983) are presented below. Data on observations made at weekly interval during one year in 1983 (*kharif*) and 5 years data of monthly distribution of two major crop pest species (*H. armigera* and *S. litura*) are presented in Tables 8 and 9.

Light Trap Catch in Relation to Prediction of Infestation Level

Light trap data available since 1977-78 to 1984-85 were considered comparable. Overall degree of infestation level of a season (Dec. to Feb.) of gram pod borer *Heliothis armigera* on chickpea crop was correlated with light trap catches obtained during September to January on monthly basis (Table 10).

Table 8: Weekly Distribution of Major Insect Pest Species Collected on Light Trap at Jabalpur during 1981-82 *Kharif* Seasons (Light source: MV 160 watt.)

Period (Weekly Interval)	Mean Catch per Night					
	Hyblaea puera	Heliothis armigera	Spodoptera litura	Agrotis ipsilon	Plusia chalcytes	
July 27	1.5	2.5	0	0	1.5	
28	4.0	2.5	0	0	2.7	
29	47	0.7	5.5	0	54	
30	100	15	32.2	0	119	
31	1.0	5.2	9.8	0	102	
Aug. 32	20	3.8	4.8	0.1	62	
33	114	4.0	10.7	0.5	98	
34	16.5	4.7	24.7	0	212	
35	2	8.0	27.8	0.1	153	
Sept. 36	0.5	6.5	2.1	0.8	65.5	
37	0.7	1.5	13.2	1	113	
38	0.8	0.7	55.1	0.5	118.5	
39	0	3.2	92.2	1.2	90	
Oct. 40	0	3.5	49	0.2	48	
41	0	3.0	8.5	0	10	
42	0	1.7	3.0	0	15.2	
43	0	1.2	–	0	11.6	
44	0	0.8	10.4	0	20.4	

Table 9: Monthly Distribution of Light Trap Catches of *Heliothis armigera* and *Spodoptera litura* Recorded in different Years at Jabalpur

Total Monthly Catch/Trap (Corrected to 30 days) in different Years

Period	Heliothis armigera					Spodoptera litura				
	85-86	86-87	87-88	94-95	95-96	82-83	83-84	84-85	85-86	86-87
June	2	—	2	—	—	145	04	31	00	—
July	3	8	7	—	—	05	45	43	22	30
August	13	8	33	—	—	102	373	898	112	126
September	12	113	102	9	1	480	345	707	679	546
October	2	47	26	32	6	965	343	713	224	515
November	2	5	22	10	17	211	183	69	118	94
December	115	55	72	66	423	234	30	34	34	19
January	162	145	887	71	297	106	95	28	273	35
February	660	1189	955	245	1452	91	298	58	359	106
March	2579	1882	3101	6493	4189	183	458	225	320	134
April	1133	12574	450	5658	3370	78	371	141	149	297
May	—	265	90	1340	82	30	146	172	—	144

Table 10: Overall Relationship between the Level of Light Trap Catch of Adults and Infestation Index of *H. armigera* Larvae

Observation Year	Infestation Index Value	Light Trap Catch (Monthly)		
		Sept. Only	Sept. and Oct. Total	Nov., Dec., Jan. total
	X	Y_1	Y_2	Y_3
1977-78	7.0	134	167	717
1987-88	5.5	102	128	981
1981-82	5.2	82	–	2533
1983-84	4.9	42	84	595
1980-81	2.8	125	404	427
1985-86	2.7	12	14	279
1982-83	2.2	32	80	386
1984-85	1.9	19	63	258
'r' value		**0.655**	**0.396**	**0.503**

Ref. Vaishampayan (1995)

Results revealed that catches were distinctly higher in September during the years of high infestation of 1977-78 and 1987-88 (134 and 102 moths) compared to years of low infestation of 1985-86, 82-83 and 84-85 (12 to 32 moths only). In October also the trend was similar. Correlation coefficient ('r' value) was highest when infestation was compared with September catch (0.655).

Direct Control of Pest Population

In recent years use of light trap has gained widespread importance in integrated pest management (IPM) strategies against potential pest species doing considerable damage in various crops. Light trap works as pest management tool indirectly through monitoring the adult activity, which decides the need of pesticide application and its timing if necessary. And directly by mass trapping of adults (male and female both) and disrupting reproduction and oviposition. Mass trapping minimizes infestation in field significantly without any toxic hazards.

From an agricultural point of view the use of light traps in pest control is far more convenient than the indiscriminate use of insecticides, because majority of useful insects (Parasites, predators and honey bees etc.) being diurnal, are less attracted by light trap and are spared (Mezokhin Porshnyakov, 1969).

Taylor and Deay (1950) observed 57 to 63 per cent reduction in the population of corn borer (*Heliothis zea*). Several workers in the United States reported significant reduction in the infestation of tobacco hornworm *Menduca sexta*, using light traps, at a density of 3 to 6 traps per square mile (Lawson *et al.*, 1963, Gentry *et al.*1967, Lam *et al*, 1968 and Cantelo *et al.*, 1972). Hamilton and Steiner (1939) observed significant reduction in the infestation of Codling moth (*Cydia* sp.) in apples using light trap. Light trap with 15-watt black light lamp was found to be most effective against pink bollworm moths *Pectinophora gossypiella* and other insects in cotton crop (Glick and Hollingsworth 1955).

Patel *et al.* (1981) from Gujarat observed the use of light traps being very effective in controlling the population of "kutra" (*Amsacta* spp.) and recommended to install the traps in the whole area on cooperative basis, soon after the first shower of monsoon rains to control the pest easily at nominal cost. Results of field trials on survey and control of insect pest species of economic importance at global level have been discussed in detail in chapter 10.

It has a tremendous potential as IPM tool in the management of bollworms, cutworms, hairy caterpillars, stem borers (sugarcane and paddy), *Spodoptera*, Diamond back moth, fruit sucking moths, codling moth, white grubs, field crickets, grasshoppers, and against paddy pests including plant hoppers, ga;;midges, gundhi bugs etc.

Light traps have a specific role against insect pest of crops grown in *Kharif* season (starting from mid June till October) specially in crops like paddy, soya bean, groundnut and sugarcane. These crops suffer with extensive damage caused by variety of insect pests belonging to order Lepidoptera, Coleoptera, Hemiptera and Diptera.

During monsoon season, most of the pesticides applied on crops are washed away due to frequent rains. The light trap on the other hand continues catching harmful insects, pests without any interruption.

In certain fruit crops like apples and grapes where acceptable level of pesticide residues (for export market) is very low or in organic farming and mushroom cultivation in rooms where use of pesticides is prohibited use of light trap is the best alternative to minimize the pest damage.

Black Light trap (with UV lamps) have a very specific role in forest areas and plantation of forest trees of economic importance where pesticide application is not practical in large areas. Another advantage of UV lamp is that UV light does not disturb the wild life because it is a dark light for them while bright MV light may disturb their activity in natural forest. It has tremendous potential against the wood borers in sal and Salai forests and plantations of teak and eucalyptus and white grubs in teak nurseries.

Observations on Control of *Agrotis ipsilon* and *Heliothis armigera* on Gram Crop and *Spodoptera litura* on Soybean Crop using Light Traps

Excellent results were obtained in controlling population of cutworms *Agrotis ipsilon* in about 20 hectares of chickpea crop in village Kathonda (Vaishampayan, 2009). Four traps, using 160-watt mercury vapour lamp, were operated from 15th of October till maturity of crop. Comparative data of incidence of cutworms in the endemic area, recorded for 9 years since 1974-75, presented in Table 11 demonstrates very significant reduction in cutworms' damage, requiring no insecticidal treatment.

In case of *Heliothis armigera*, 5 years data presented in Table 12 -revealed 50 to 60 per cent reduction in pest population in chickpea crop around light trap within

Table 11: Relationship between Rainfall, Light Trap Catches and the Incidence of Black Cutworm, *Agrotis ipsilon* at Jabalpur

Year of Observation	Total Rainfall (mm)		Total Catch/Trap in Non-endemic Area (Total of Nov. and Dec.)	Cumulative Trap Catch Octo. to Dec. in Endemic Area	Incidence of Pest in Dec./Jan Larval Population/m Row (max. weekly mean)
	Monsoon Season July to Oct.	Winter Season Nov to Dec.			
1974-75	747	2.0	265+	Nil	3.5
1975-76	1238	0.0	35+	Nil	3.8
1976-77	918	8.2	78	Nil	7.0
1977-78	1424	30.5	64	Nil	9.6
1978-79	921	0.0	57+	Nil	4.8
1981-82	465	42.8	55	Nil	0.4
1982-83	1221	17.1	21.3	883	3.0**
1983-84	1252	8.4	65	465*	3.3**
1984-85	1284	0.0	174	5096	1.0**

* December only * Late installation of light trap. 100 per cent losses in February.

** 1.During 1983-84 and 84-85 very high incidence of pest was anticipated (6-8 larvae per meter).

2. Light trap operation help to reduce the incidence very significantly. Over 80 per cent reductions in pest incidence was observed in 1984-85 crop season.

a radius of about 100 meters with a population level of 9 to 15 larvae compared to 17 to 19 larvae per 10 samples 200 meter away from the trap.

Table 12: Larval population of *H. armigera* on Gram Crop at Various Distances from Light Trap (mean of four years, 1982 to 1986)

Month/Week	Larval Population/10 Samples (1/2 x 1/2 m crop area)			
	50 m	*100 m*	*150 m*	*200 m*
December I	5.0	5.0	5.0	21.0
II	–	–	–	–
III	–	–	–	–
IV	1.6	5.6	1.7	0.5
January I	2.5	7.5	6.2	11.2
II	2.6	4.2	4.5	5.8
III	2.9	10.2	7.1	4.5
IV	7.8	9.8	7.3	6.9
February I	14.2	12.8	18.1	15.3
II	19.9	25.6	20.2	16.6
III	8.5	21.5	14.5	13.5
IV	14.7	21.3	22.4	17.6
March I	12.2	16.5	27.2	14.0
II	2.0	13.0	7.0	9.0
III	22.0	40.0	7.5	85.0
Mean	**8.90**	**14.85**	**16.63**	**18.9**

Observation on Control of Spodoptera Larvae on Soybean Crop

Observations were recorded on larval population of *Spodoptera litura* larvae at various distances from light trap during 1983-84 crop seasons. Abstract data are presented in Table 13.

Results show the larval population (infestation level)

  (i) Close to trap from 50-150 mt. was 1.3 to 5.3 larvae/10 sample.

  (ii) 200 mt. away incidence was 8.0 larvae/10 samples.

  (iii) In non trap area 1 km away- incidence was 11.3 larvae/10 samples.

In conclusion light trap operation reduced the infestation very significantly. The control was over 60 per cent of larval population.

Observations on Use of Light Trap in the Management of Red Hairy Caterpillar *Amsacta moorei*

A. Work done in Madhya Pradesh

Extensive work was done during the years 1982-1986 to test the use of light

Table 13: Larval Population of *Spodoptera litura* on Soybean at Various Distances from Light Trap (Mean of two years–1983 and 1984)

Period Weekly	Larval Population/10 Sample*				
	Distances from the Trap				
	1 km away**	200m	150m	100m	50m
July					
IV	53.0			12.0	0.0
August					
I	10.0	4.0	2.0	24.0	2.0
II	2.5	35.0	7.0	7.0	3.0
III	4.5	13.5	8.5	5.0	6.0
IV	1.0	1.0	1.0	0.0	0.5
September					
I	1.0	0.0	0.0	0.0	0.0
II	19.0	1.0	2.0	0.0	0.0
III	0.0	0.0	0.0	0.0	0.0
IV	0.0	0.0	0.0	0.0	0.0
Mean of a season	**11.3**	**8.0**	**3.3**	**5.3**	**1.3**

*Sample unit (1/2x1/2 meter)** non trap area.

trap in survey and control of *Amsacta moorei*, a polyphagous pest of *Kharif* crops in the endemic area of district Jhabua, Madhya Pradesh (Vaishampayan, 2009).

Light traps of JNKVV model with 160-Watt Mercury Vapour lamps were operated every night during the active season and trap caches were examined every day. Observations on larval population on major host and trap crops were taken following standard sampling procedures. Data are presented in Tables 14 and 15.

Results of significant importance are summarized below.

1. Pest is active from mid June to August.
2. Moth emergence and collection of moths in the traps starts in significant numbers on 3rd and 4th day after the first monsoon showers with precipitation above 50 mm within 2 days. Egg laying is followed immediately. Relative size of trap catches helps to predict areas of low as well as high infestation.
3. Female adults pass through single or multiple mating by male adults prior to oviposition which enables them to fertile eggs to continue future generation. Trapping of these male adults in large number adversely affect the fertility status of gravid females and helps to reduce the laying of fertile eggs.
4. Second generation larvae emerged in August or September, pupate in soil and undergo long diapauses period of about 10 months in pupal

Table 14: Weekly Distribution of Rainfall and Light Trap Catches of *Amsacta moorei* (Arctiidae) during different Years

Total Trap Catches/Trap/Week (JNKVV trap)

Location: Jhabua

Period/(Weekly)		1983		1984		1985		1986	
		Rainfall (mm)	Trap Catch	Rainfall (mm)	Trap Catch	Rainfall (mm)	Trap Catch	Rainfall (mm)	Trap Catch
June	25	53.0	–	6.2	7	–	–	52.2	326
	26	31.0	–	0.0	2	8.0	4	10.7	6092
July	27	41.2	192	70.9	476	0.0	2	–	377
	28	40.5	322	1.7	100	16.0	1	8.2	1
	29	79.2	1107	63.0	21	7.0	165	35.5	2255
	30	53.2	567	6.0	77	27.0	16	–	–
August	31	23.0	613	–	–	50.0	37	–	–
	32	119.2	2282	–	–	19.0	18	–	–
	33	66.0	1050	–	–	15.0	930	–	–
	34	12.7	507	–	–	11.0	418	0	0
	35	0.0	–	–	–	–	16	–	–
Total		**519.2**	**4660**	**147.7**	**683**	**153.0**	**1607**	**85.0**	**9051**

Table 15: Survey of the Incidence and Control of Red Hairy Caterpillar *Amsacta moorei* in Jhabua Distt. July 1984

Sl.No.	Location	Crop Observed	Number of Larvae per 10 Samples*	Remarks
A.	**JHABUA Block:**			
(i)	Gopalpur	Maize	20 larvae	Control**
(ii)	Dhekal	Maize	5 larvae	Light trap installed
(iii)	Dhekal	Maize	2 larvae	One year operation
B.	**BHABRA Block:**			
(i)	Kakarbari	Maize and Urd	14 larvae	Control**
C.	**JOBAT Block:**			
(i)	Kilajobat	Cowpea, Maize	30 larvae	Control**
		Maize	19 larvae	
		Cowpea, Moong	49 larvae	
		Cowpea, Moong	34 larvae	
(ii)	Kalikhaetar	Sunhemp	1 larvae	Light trap installed in 1984
		Maize	1 larvae	(One year operation)
		Cowpea	0 larvae	
(iii)	Jobat farm	Moong	1 larvae	Light trap installed in 1982
		Urd	0 larvae	(Three years operation)
		Urd	0 larvae	

* Sample Unit- ½ x ½ meter **Control means "No light trap operation"

stage. Trapping of these adults during August/September will check the carryover of the pest and reduce its activity in the following year.

5. Over 80 per cent reduction in larval population was observed in field crops close to light trap, covering around 50 hectares of crop area. Reduction was absolute 100 per cent on Jobat farm in about 20 hectares, where light trap was in operation consecutively for 3 years since 1982.

B. Work Done in Gujarat

Patel *et al.* (1987) conducted studies on the management of Gujarat hairy caterpillar, *Amsacta moorei* (But.) using ultra violet light trap on groundnut crop in Gujarat, India, during 1981-83. The traps were installed at the rate of 11 per 10 ha, 7 per 8 ha and 8 per 10 ha in 1981, 1982 and 1983, respectively. An area about 30 ft in diameter around each trap was dusted with 5 per cent BHC (HCH) to kill hatching larvae. The first adults were trapped 5, 3 and 7 days after the monsoon showers in 1981, 1982 and 1983, respectively. In 1981 and 1982, 98.7 and 84.1 per cent, respectively, of the total catch over 12 days was made in the first 5 days, while in 1983 73.6 per cent of the total catch was made in the first 6 days. The mean larval population was 20.42-72.25 and 144.5-396.6 individuals/300 ft^2 in areas with and without light traps, respectively, in 1981. The larval population in 1982 was drastically lowered at 8.49-11.9 and 18.99-24.66 individuals/300 ft^2; this was a result of the large catch in 1981.

C. Observations of S. Pradhan (1969) on Control of Red Hairy Caterpillars

The pest spends the whole of winter right from September and early summer in the pupal stage in the soil up to a depth of about half a meter. They emerge from the soil during the monsoon in several successive waves, the first of which comes up soon after the onset of the rains. These adults are attracted to light. Therefore, light trap should be set up as soon as rains start. The light traps will serve a double purpose: firstly, they will kill quite a large number of Kutra moths (Kamlia keet) before they are able to lay eggs and, secondly, they will provide a warning about moth emergence from the soil.

Light traps have been recommended for a very long time but they have never been used as definite items of a well organized multi-pronged campaign. Keeping in view the peculiar life-economy of the kutra pest, will convince anybody that a light trap is bound to be a very effective item of the kutra campaign. Hence, in areas where kutra trouble appears year after year the light trap should constitute a normal piece of equipment which the farmers should keep like a plough, seed drill etc.

9

General Types of Electric Insect Traps

Currently designed insect traps that utilize an electric lamp as the attractant are frequently classified into three types or groups: (1) the electrocutor or electric grid. (2) the suction- or fan-type trap; and (3) the gravity or mechanical trap.

(1) Electrocutor or Electric Grid-Type Trap

The electrocutor or electric grid-type of light trap consists of a series of parallel wires, with adjacent wires insulated from each other and connected to opposite polarities of a low-current high-voltage transformer. The electrocutor-type insect trap is commonly used for insect destruction purposes. It is widely used in attempts to eliminate or control nuisance insects around residences, business establishments, and food processing plants. The arc action between grid wires may cause burning or other bodily damage to insects passing between the wires making insect identification difficult. For that reason, this type of trap is seldom used for insect surveys.

(2) Suction- or Fan-Type Trap

Suction- or fan-type electric insect traps were developed to fill the need for catching small, agile fliers which tend to hover near the lamp or alight on the baffles. Herms and Bergess (1928) added a suction device in addition to the attractant lamp. A suitably designed, small electric fan was selected for that purpose. A modification of this original trap, known as the Akins trap, was reported by Essig (1930).

The Akins' trap consisted of the following parts: (1) An inverted, funnel-shaped reflector, 18 inches in diameter and provided with a light socket; (2) a 100-W lamp to attract the gnats; (3) a thin sheet iron sleeve, 15 inches in diameter, suspended about 10 inches from the reflector by three ¼ inch rods, allowing clearance for the gnats to come to this light; (4) a small electric fan with motor, the latter attached to the sleeve and designed to draw the insects down from the light by the action of the fan; (5) a black muslin bag, about 3 feet long, drawn over the lower end of the sleeve (Figure 23).

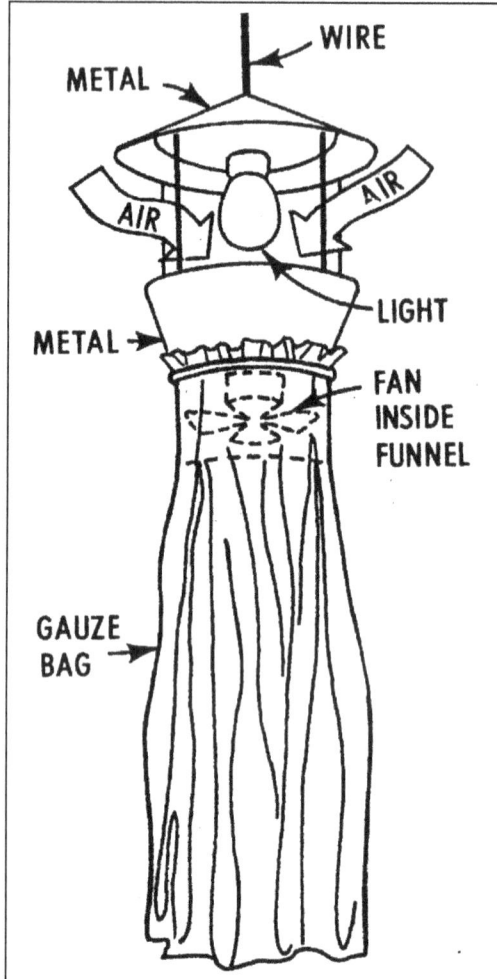

Figure 23: The Akins Gnat Trap (Suction trap).

(3) Gravity or Mechanical Trap

The gravity or mechanical trap is the third type of insect trap that uses an electric lamp as the attractant. It differs from the electrocutor type in that it has no high voltage for killing insects and from the suction-type trap in that it has no motor-driven fan. Thus, its capture of insects depends entirely on the attractant lamp and the trap design. Insect collectors have used the gravity type most often because the insects captured are less damaged and more easily identified. Most of the light traps models used in insect surveys and pest control studies referred herewith and those developed at JNKVV Jabalpur are of gravity or mechanical type trap.

Development of Light Traps for Insect Surveys

Walkden (1942) pointed out that "light traps offer an efficient means of obtaining information regarding the distribution, seasonal flight periods, and peaks of abundance of various insect species. Taylor and Deay (1950) made laboratory tests in 1948 on the relative attractiveness of various gaseous discharge lamps to the European corn borer moth and concluded that the maximum attractiveness for this moth (at the intensity levels studied) was in the near ultraviolet region between 320 and 380 nm. They also reported that among other insects attracted were the adults of the tomato hornworm, *Manduca quinquemaculta* (Haworth), the tobacco hornworm, *Manduca sexta* (Johannson), and the corn earworm, which were attracted in great numbers to sources radiating in the near ultraviolet.

Omnidirectional Trap

Changes in design of the electric insect trap with the vertically mounted BL lamps were developed from study of its performance and suitability in trapping insects. Results of the comparisons in 1955 by Hollingsworth and Carter

Figure 24: Omnidirectional Light Trap, Developed in Texas with One 15-W BL Lamp.

(unpublished) showed that the omnidirectional trap with one 15-W BL lamp (Figure 24) collected the greatest number of pink bollworm moths and all other insects. The mercury vapor trap (with 100-w. MV lamp) caught second largest number of all insects and the same number of pink bollworm moths as the unidirectional trap.

Development of BL Trap Standards (USDA)

In 1964, the Committee on Insect Surveys and Losses, Eastern Branch, Entomological Society of America (E.S.A.) was formed to take action for the development of BL trap standards. A questionnaire was sent to 53 survey and extension entomologists, agricultural engineers, and plant pest control people known to be actively engaged in insect light-trapping research or service programs for States and USDA. Thirty-six questionnaires were returned to the committee, and 21 of them indicated that general insect survey BL traps should be standardized. A study of the returned questionnaires indicated that a majority of questioners recommended the use standard light trap equipped with one linear 15-w. fluorescent BL lamp. A majority recommended the use of an omnidirectional trap with baffles, no fan, and a collection funnel opening (top) with a 10- to 18-inch diameter.

On the basis of these reviews, the Committee proposed that certain trap design standards be used in general insect survey work. The design should include: One 15-W BL lamp; omnidirectional design; baffles; no fan; 10- to 18-inch funnel opening (top diameter); and a collection container larger than 1-quart (Figure 25). The specifications of the trap dimensions and standards are described below.

Standards

1. Attractant-one F15T8/BL lamp (15-watt black light) mounted vertically. See A.
2. Position of lamp-bottom even with rim of funnel, lower lamp holder below rim.
3. Four baffles (two crossed), dimensions: total width 14"; total height above funnel rim 19"-20"; clearance between inner edge and lamp ¼ "- ½ "- See B.
4. Funnel-slope 60°; top diameter 14" (approx. ¾ length of lamp); bottom opening diameter 2"; lower end inserted into top of collection container ¼ " to form drip rim for water. See C.
5. No large canopy over top of baffles (such a cover reduces catches of some species).

Additional Recommendations

1. Wiring system showing Underwriters' Laboratories (UL) seal of approval.
2. Electrical components mounted either one side or top, but if on top the area of obstruction to light not to exceed a 5" square (25 sq. in.).
3. Use of a side-emptying drain placed in cover of collection can to leave collection container unobstructed. See D. Pan diameter at least 4", depth 1", drain opening ½ " x 1 " minimum.

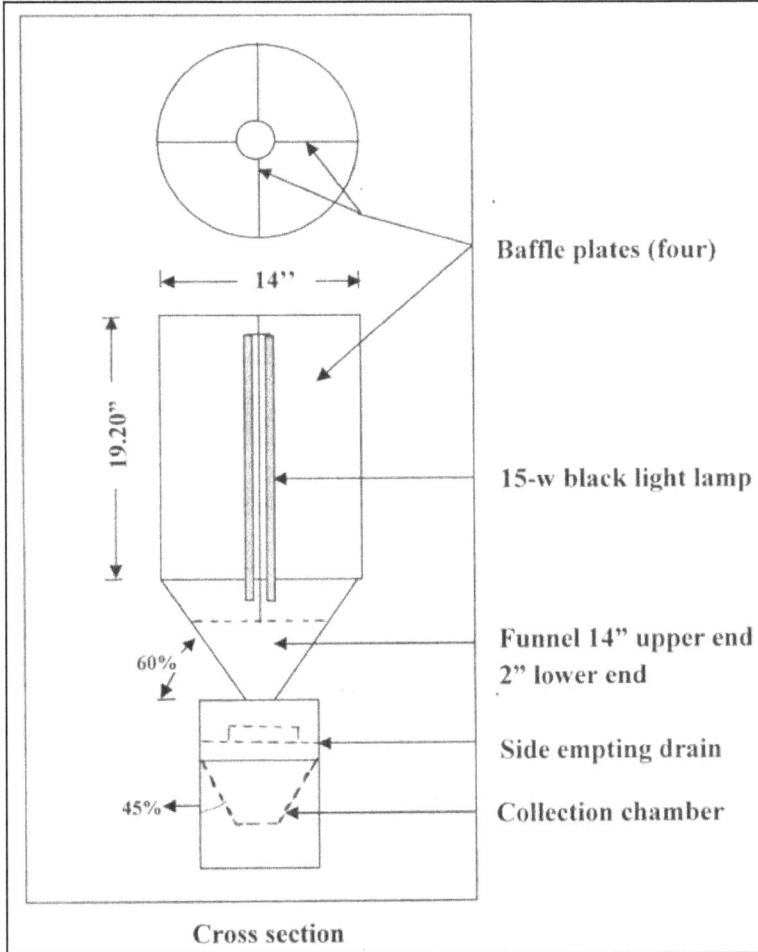

Figure 25: Black Light Trap Standards for General Insect Surveys (USDA, 1964).

4. Collection container designed for use of less hazardous killing agents, such as ethyl acetate (as compared to calcium cyanide) through use of insert funnel with sealing gasket, 45° slope and 2" opening. See E.
5. Material—26 gauge galvanized steel minimum.

Survey Traps for European *Chafer* Beetle

Adults of the European *chafer, Amphimallon majalis* (Razoumowsky), were strongly attracted to radiation from fluorescent BL lamps tested by Tashiro and Tuttle (1959). When exposed to very low populations late in the 1958 season, traps containing these lamps captured up to 70 times as many beetles as the most attractive, chemically baited traps. Studies of factors involved in the design and development of an effective survey trap for the European chafer were conducted by Tashiro, Hartsock, and Rohwer (1967) during the period 1959-63. This survey trap is suitable for either European chafer beetle surveys or general insect population survey.

Hardwick's Noctuid Light Trap

Hardwick (1968) while presenting a brief review of the principles of light trap designs has described a design of an efficient trap- a noctuid trap. This light trap as illustrated in Figures 26A and B has been used with good results for collecting noctuid moths during several years of survey work. Trap is easy to install and operate keeping the collection free from rain water.

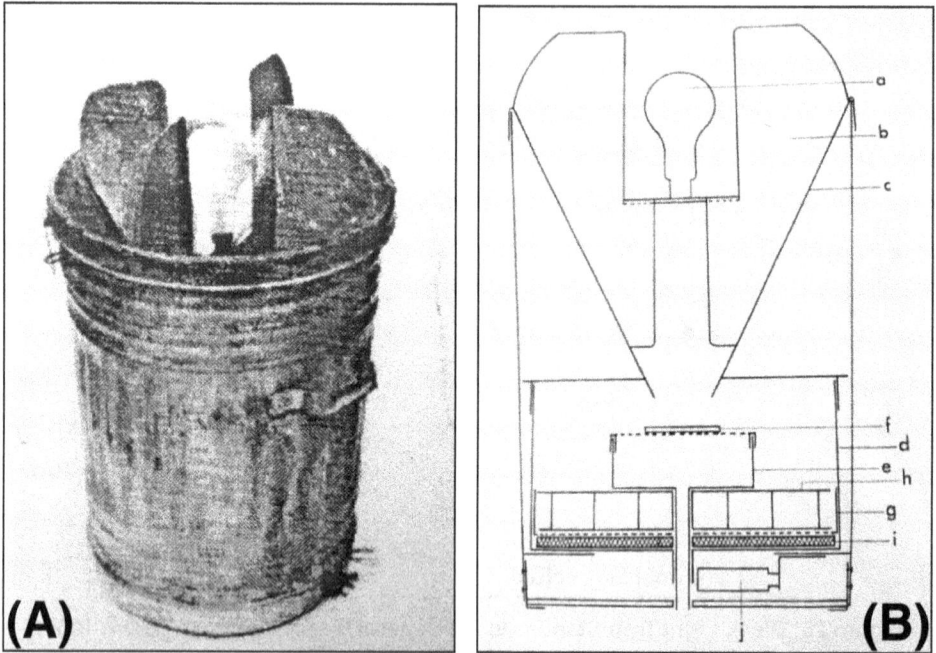

Figure 26: Hardwick's Noctuid Light trap Exterior View (A) and Vertical Section (B).
a: Light source; b: Baffle; c: Cunnel; d: Inner metal chamber; e: Screen-lidded rain drain; f: Disc of sponge rubber; g: Removable metal tray; h: Removable metal lattice; i: Cheese cloth pad; j: Heating element.

Light Trap Models Developed by Vaishampayan S.M. (1982-2014)

1) Trap Designs for Scientific Studies

All the models developed at JNKVV, Jabalpur, Madhya Pradesh (SM- 82, SM-85, Pannsylvanian, SM-88 and SM-01 as shown in Figures 27–29). These models are suitable for scientific studies only, maintaining its trapping efficiency at highest

LIGHT TRAP MODELS DEVELOPED BY S.M. VAISHAMPAYAN AT J.N.K.V.V. JABALPUR (M.P.)

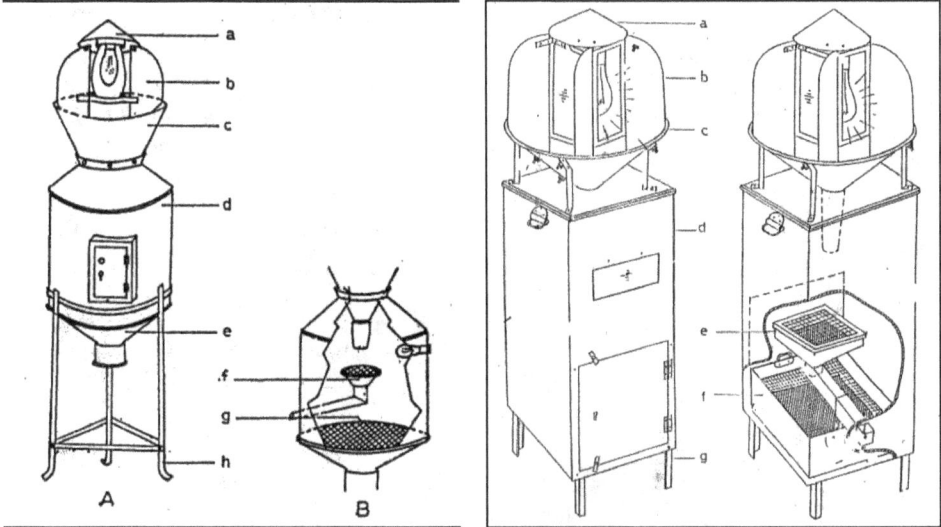

Light Trap Model SM-82
(Vaishampayan, 1985)

Light Trap Model SM-85
(Vaishampayan, 1982)

Figure 27

a: Rain shade with bulb and glass cage; b: Baffle plate; c: Funnel; d: Collection chamber with door; f/e: Rain drain system; g/f: Wire mesh screen bottom/collection trays; h/g: Stand.

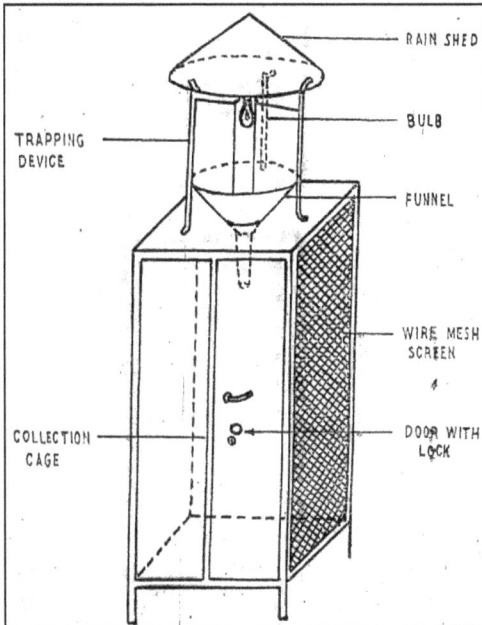

Figure 28: Modified Pennsylvanian Trap (Vaishampayan, 1980).

Light Trap Model SM-88   **Light Trap Model SM-01**

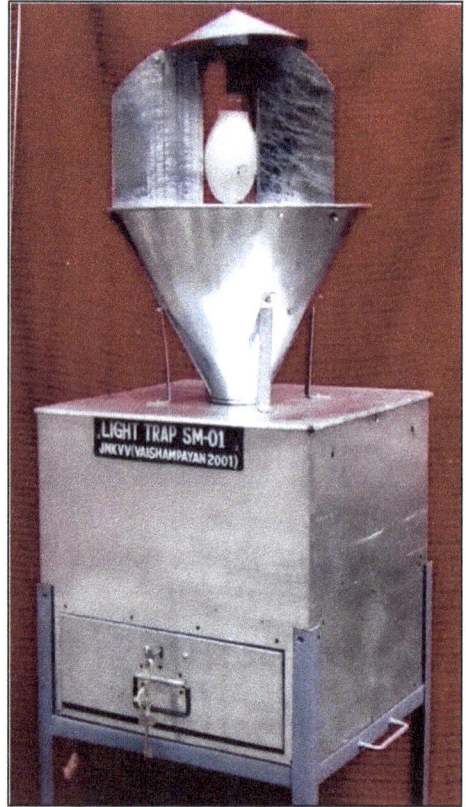

Figure 29: Light Trap Designs Developed by Vaishampayan in 1988 and 2001.

level (quantitative aspect), and preserving this specimen in good condition, safe from rain water (qualitative aspect).

Trap model SM-82 was installed at government farm Jhabua, MP and successfully operated for 3 years during 1983-85 for experimental work against red hairy caterpillar, *Amsacta moorei*, a major pest in *Kharif* season. Pest was eradicated in 20 acres of area after 3 years. A record collection of single day's trap catch with total biomass (all insects) of 8.5 kg in weight was obtained on 24th September 1981(Vaishampayan, 1982).

2) Trap Designs for Mass Production in Plastic Mould and Large Scale Use in Farmer's Fields

(i) NCIPM (ICAR) Model of Trap (2010)

This trap was designed by National Center for Integrated Pest Management (ICAR) New Delhi and produced by Fine Trap (India) Yawatmal, Maharashtra

 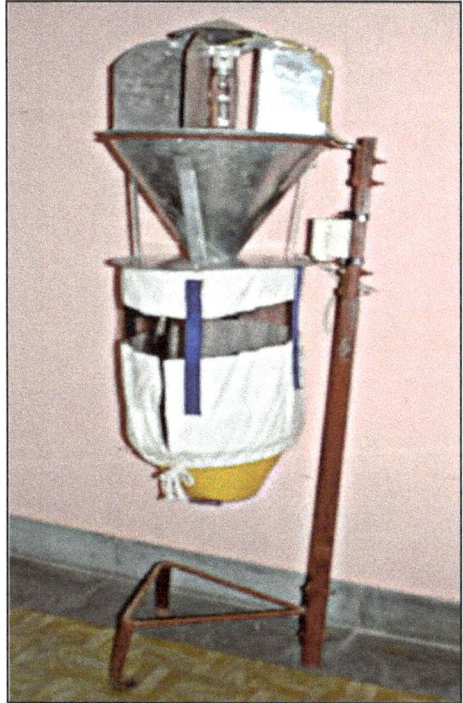

SMV-4 (UV)   SMV-4 (MV)

Figure 30: New Designs of Light Trap Developed by Vaishampayan in 2014.

(See Figure 31). This trap was claimed to be eco-friendly permitting the beneficial predators and parasites to escape. There were some shortcomings in the model, hence it was redesigned.

(ii) Redesigned NCIPM (ICAR) Model in Collaboration with Fine Trap (India)

Model SMV-3 (M.V. and U.V) (2014)

Several changes have been made in redesigning, bringing improvement in trapping efficiency as well as in preserving the collected specimen in good condition and easy removal of trap catch from the collection box. (See Figure 32)

Model SMV-4 (M.V. and U.V.) (2014)

In these models, the insect collection box (made of G.I sheet) is replaced by folding collection bag made of soft nylon or teryline cloth or soft rexin sheet provided with 8" long chain for removal of live specimen of insects if needed. Length of baffle plates above the funnel rim is 9" in M.V. model and 12" in U.V. model (see Figure 30). For line drawing and dimensions see Figure 32.

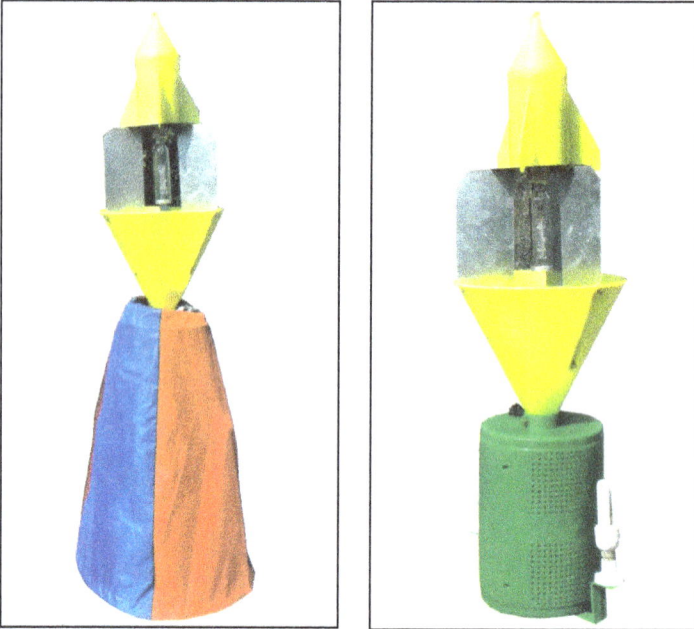

Figure 31: Light Trap Designed by NCIPM (ICAR) New Delhi in 2010 and Produced by Fine Trap (India) Yawatmal (Maharashtra).

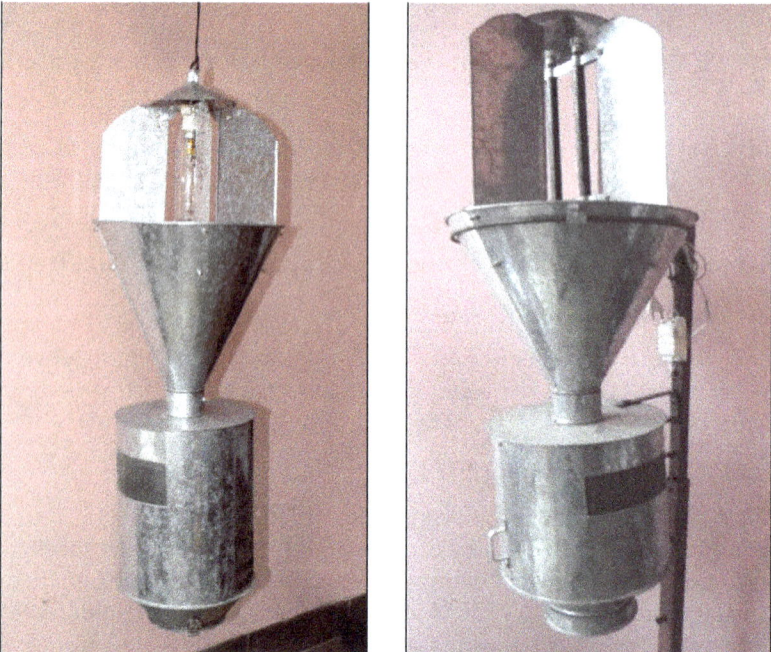

SMV-3 (MV)   SMV-3 (UV)

Figure 32: Redesigned NCIPM Model Developed by Vaishampayan in 2014 in Collaboration with Fine Trap (India) Yawatmal (Maharashtra).

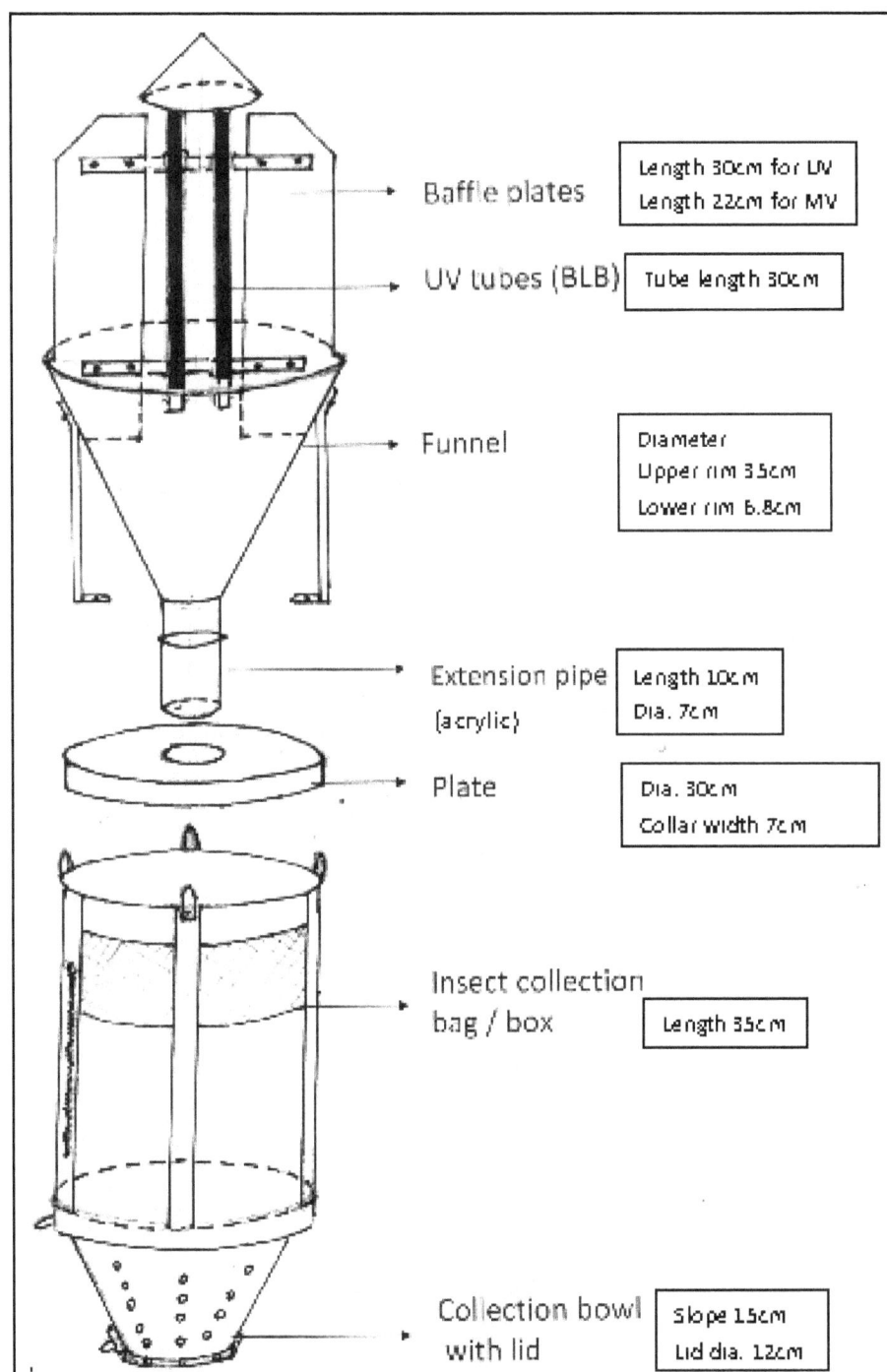

Baffle plates
| Length 30cm for UV |
| Length 22cm for MV |

UV tubes (BLB)
| Tube length 30cm |

Funnel
| Diameter |
| Upper rim 35cm |
| Lower rim 6.8cm |

Extension pipe (acrylic)
| Length 10cm |
| Dia. 7cm |

Plate
| Dia. 30cm |
| Collar width 7cm |

Insect collection bag / box
| Length 35cm |

Collection bowl with lid
| Slope 15cm |
| Lid dia. 12cm |

Figure 33: Line Sketch of Trap Models SMV-3 and SMV-4 (MV and UV).

10
Role of Light Trap in Study of Insect Migration

Light trap serves as an important instrument in monitoring the migratory movements of insect population. Analysis of certain characters morphological and physiological provides evidences to identify a migratory phase of a population *i.e.* immigrant, resident or locally bred and emigrant and its consequent effect on further breeding of a pest in the area. Since migration in insects is closely linked with the population dynamics of a pest species including the occurance of outbreak of population, the concept of migration in insects and its role in population dynamics of a pest species and the basis of identification of various migratory phases, along with experimental evidences etc. are discussed in detail below.

Concept of Migration in Insects

Migrations in insects are adapted for the survival of the species and they are linked with almost every other aspect of insect life, especially reproduction. Migrants show explosive reproductive pattern (*i.e.* they reproduce rapidly), which serve to divert additional energy to egg production, their ability to exploit newly colonized sites is known to influence the built up of the population of larval or nymphal stages quite significantly in an agroecosystem. This special ontogenetic characteristic of mass movement from one breeding site to others is a part of the chronological sequence in the life history of the genotypical migrants (Wigglesworth 1963 a, b., Johnson 1963 b, Rainey 1965).

Migration is often linked with the need of a population to relinquish a habitat that is only transiently fit for breeding (Southwood, 1962) and the same is common in many of the noctuid Lepidoptera. Earlier, it was believed that the migratory flight of the insect was due to its response to currently adverse factors. Kennedy (1956) and Johnson (1963 a) however, postulated that particular unfavorable factors in an environment act as token stimuli, influence ontogeny and cause insects to develop into migrants with involvement of the endocrine system. They associated migration with immaturity of ovaries, development of broad wings with strong wing-muscles and fat-accumulation, all due to deficient corpus allatum secretion and further revealed that migration ceases when ovaries become mature with active corpus allatum. The phenomenon of immature ovaries associated with mass migration was discovered as early as 1911 by Pospielov while working on *Agrotis segetum*. Johnson (1963a) found that among migrants, females are more significant and predominant as their conditions are more regularized. It is their function to Iay eggs in suitable places while fertilization may be done enroute or at journey's end.

Migration in *Heliothis armigera*

Farrow and Daly (1987) defined three categories of movement by *Heliothis* spp. *i.e.* short range long range and migratory movements. All are important in colonization and exploitation of agroecosystem as defined below.

(i) Short range of trivival movement occurs within the host canopy and involves appetitive behaviors such as feeding, oviposition and mating and is confined to 100-1000 meter.

(ii) Long range movement from 1 to 10 km, occurs about 10m above the canopy and usually downwind (Drake and Farrow, 1988,; Farrow and Daly, 1987). Long range flight includes movement between crops and between the feeding and oviposition sites and is classed as migratory.

(iii) Migratory movement occurs above flight boundary layer at an altitude of 1-2 km and may continue for several hours.

Heliothis spp. are facultative migrants and migrate in response to poor local conditions for reproduction *i.e.* short of adult nectar sources or larval food, and the passage of weather conditions conducive to such movements.

Mark-recapture experiments confirmed movements of moths over a distance ranging from 25 to 160 km (Hendricks *et al.*, 1973). Movement of *H. zea* from southern Texas to south-central Arkansas covering distance from 75 to 1000 km has also been documented earlier (Hendricks *et al.*, 1987). Pedgley (1985) observed that *H. armigera* migrated up to 1000 km to reach Britain and other parts of Europe from sources in southern Europe and North Africa.

Radar plumes have been observed arising from the fields of groundnut in Sudan and from cotton fields in Arizona (USA) after the take off flights starting 50 to 60 minutes after sunset (Lingren and Wolf 1982; Schaefer 1976)).

Analysis of 3 years' data on light trap catches of *H. armigera* adults collected at Jabalpur (23°N) indicated migratory nature of the species distinguishing the population into three distinct migratory phases *viz.* (i) Immigrants during December,

(ii) locally bred-residents during January and February and (iii) emigrants during late February and March (Vaishampayan and Verma, 1987).

Migration in *Agrotis ipsilon*

It is a pest of major significance of rabi crops in Madhya Pradesh, Uttar Pradesh, Punjab, Rajasthan and especially in Bihar. In Bihar, this pest causes very extensive damage, where nearly 20-25 thousand hectares area is infested annually (Singh, 1971).

Johnson (1969) classified migrant insect into three broad classes, *Agrotis ipsilon* is classified in the 3rd type where long-lived, sexually immature adult insects migrate from breeding-site to diapauses-site and next season retrace the emigration route to return to the same original breeding-site. These millions of noctuids make mass migration over long distances, sometimes exceeding the distance of 3,500 kms (Mikola and Salmensua, 1969).

Williams (1926) reported migration in the 3rd brood of the pest which emerges in March and April with greatly developed fat body and immature ovaries. In 1937, he classified *Agrotis* as a regular long distance migrant from Egypt to Europe. Johnson (1969) reported probable migration of *Agrotis ipsilon* towards the Himalayas some 360 km from the Ganga valley.

Many workers had considered that *Agrotis ipsilon* migrates in hilly regions during summer and comes back to plains during the active season to re-infest the areas (Fletcher, 1916; Williams, 1926; 1937; Davis, 1953; Kapoor, 1955; Malicky, 1967; Novak and Spitzer, 1972; Odiyo, 1975; Oku and Kobayashi, 1978; Nasar and Hosny, 1980 and Vaishampayan, 1981).

Role of Migration in Population Dynamics

Adults occupy an important position in the life cycle of the insect and influence its population dynamics very significantly. On depletion of principal food source, increase in population pressure or prevalence of adverse condition the female population decides to discontinue breeding in the same locality and moves to other place in the interest of the progeny. This they achieve through physiological control over maturation of ovaries, cesation of mating and provisioning of food reserve (fat bodies) and ultimately switch over to new breeding site en masse as migrants (Johnson 1969; Vaishampayan and Verma 1987).Such movements influence the population development at both the ends; depletion at source (emigrtion-outward) and addition at destination (immigration-inward).

The concept of migration in noctuid Lepidoptera as a major factor influencing the population dynamics of the pest species has originally been developed by Vaishampayan during his constant observations made for over 10 years on the activity of many lepidopterous pest species specifically, *Heliothis armigera*, *Agrotis ipsilon* and *Hyblaea puera* collected on light trap as a part of the ICAR Project on light trap studies. Vaishampayan and Verma (1987) provided first evidence on the existence of migration in *H. armigera* in India, identifying immigrant, resident and emigrant population in light trap catches. Vaishampayan *et al.* (1987) on the basis of light trap catches and other observations established a new theory on interstate migration of Teak defoliator, *Hyblaea puera* from Kerala to Eastern. Madhya Pradesh.

Migratory species of insects of agricultural importance, specifically the noctuids are known to cause heavy damage to variety of agricultural crops at places into which they emigrate. Thousands of acres of corn and soybean were destroyed by northward movements of migrants of *Agrotis ipsilon* in Indiana USA (Davis, 1963). Similar is the case with immigrants of *Mythimna separata* in northern Japan which caused heavy damage to rice (Oku and Kabayashi, 1977). Adults of *Spodoptera exigua* have been reported to make non-stop migratory flights of 1600 to 3200 km over sea and land in 4 to 5 days (Hurst 1965 a, b, French 1964, 1966 and 1968).

South to North Movement of *Heliothis armigera*

The activity of incoming migrants needs to be monitored properly as it helps in understanding population dynamics and its consequent importance in the management of pest population. While considering factors responsible for outbreaks of migrant species knowledge of weather conditions prevailing in the habitat and also beyond the boundaries of local ecosystem from where the pest arrives as emigrant, is important (Vaishampayan *et al.*, 1987). Outbreaks of *Heliothis armigera* in Rabi season on pulses occurred in 1977-78 and 1987-88 in Central India (Madhya Pradesh) were found to be influenced by migration of *Heliothis* adults from southern states including Tamil Nadu and Andhra Pradesh where the pest breed round the year. These migrant adults leave significant impact on the buildup of early season population of *Heliothis* spp. on major host crops. The movements of cyclonic winds from south-east coastal areas of Tamil Nadu and Andhra Pradesh towards north possibly play a significant role in quick and wide spread migration of adult population of *H. armigera* to central India (Vaishampayan 1995 and 2000). Outbreaks of *Heliothis* during 1977-78 and 1987-88 seasons were found to be linked with such immigration of adults. The trap catches were extremely high at the termination of crop season during March and April but field infestation during following summer and rainy seasons remained at very low level. Reproductive analysis of the adults shows their outward migration in these seasons (Vaishampayan and Verma, 1987). The exact nature of migration behavior is however, yet to be investigated in detail by monitoring adult activity through light trap catches at national level. The circumstantial evidence in support of such a theory of migration has also come from the fact that the activity of *Heliothis* in India occurs in a gradient from South to North as shown in Table 16.

Types of Migrants

Migration plays an important role in the population dynamics of many pest species. Any migratory insect population of an area or locality is made up of following types of population.

1. Immigrants or incoming population.
2. Resident or locally bred population.
3. Emigrant and or outgoing population.
4. Migrants in transit (both ways).

Table 16: South to North Gradient of Activity (Migration) of *Heliothis armigera*

State	Latitude Range	Period of Peak Activity of Pest in Field	Major Hosts
Tamil Nadu	8°–12° N	August, September, October	Cotton, Sorghum
Andhra Pradesh	13°–20° N	October, November, December	Cotton, Pigeonpea
Maharashtra (Vidarbha)	20°–21° N	October, November	Cotton, Pigeonpea
Madhya Pradesh	21°–23° N	January, February	Chickpea, Pigeonpea
Uttar Pradesh (Kanpur)	25°–27° N	March, April	Chickpea
Haryana	29°–30°N	March, April, May, June	Chickpea, Wheat, Vegetables, Cotton,
Punjab	30°–32° N	April, May, June	Chickpea, Wheat, Vegetables

Vaishampayan, 1995.

Characteristic Features of Migratory Population

The population in different migratory phases can be identified on the basis of wingspan, ovary development, fat body content, frequency of mating and activity of the neurosecretory cells as described by Johnson (1969); Wada *et al.* (1980); Vaishampayan *et al.* (1983, 1987) and others. In order to identify different migratory phases of population it is important to know the characteristic features of migratory population which are described in brief below:

1. Migrations are always dominated by females which may or may not be accompanied by males.
2. Noctuid migrants take off after sunset. In butterflies it occurs in the morning hours.
3. Migrants have enhanced locomotors response with inhibition of 'vegetative' responses.
4. They ignore sensory inputs which will later cause them to settle, feed and reproduce.
5. Migratory flight initiation is a characteristic of young females with complete cessation of ovary development.
6. Maturation of the ovary and onset of egg production are initiated towards or at the termination of migratory flight.
7. Migratory flight is undertaken in unmated condition.
8. Mating may occur during return flight; hence alighting immigrant female populations may be dominated by single mating.
9. Mating is more frequent (more than one spermatophore in *Bursa copulatrix*) in locally bred populations.
10. Females are fully charged with fat containing cells (fat bodies) prior to initiation of migratory flight to meet the energy required for long flights.
11. Residential are dominated by adults with low fat body centonts.
12. Migrants generally have longer wingspan.
13. (i) Juvenile hormone (JH) is at low level in migratory population (emigrant).
    (ii) JH may be higher in immigrants compared to emigrants as JH titre raises at the end of migration.
    (iii) JH titre triggers the ovary development which shuts down migration.
    (iv) JH is higher in locally bred population.

Identification of Various Phases of Migratory Population

Classification of a migrant population in three categories namely Immigrant, Resident and Emigrants, based on variation in morphological and physiological characters as discussed and reviewed by Johnson (1969) and by B. Energeet (1988) are summarized in Table 17.

Table 17: Identification of Various Phases of Migratory Population of Noctuid Moths based on Analysis of Light Trap Catch

Sl.No.	Key Character	Immigrant (Incoming) (Sept.–Nov.)	Resident (Local Breeding) (Dec.–Jan.)	Emigrant (Out going) (Feb.–Apr.)
1.	Seasonal activity of Adults	Low Population	Moderate population	High population
2.	Infestation (Larval population)	Moderate	Highest	Very low
3.	Sex ratio (Female)	Low	Moderate	Highest
4.	Wing span	Medium and Long	Short and Medium	Long
5.	Body length	Medium and Long	Short and Medium	Long
6.	Flight muscles	Undergo Histolysis	Totally Hightailed (nil/low)	Fully developed
7.	Ovary condition category			
	I. Nil development	i. Absent	i. Very low	i. Highest
	II. Little development	ii. High	ii. Very low	ii. High
	III. Moderate development	iii. High	iii. Low	iii. Low
	IV. Fully developed	iv. Low	iv. High	iv. Low
8.	Mating frequency category			
	I. Virgin (unmated)	i. Rare	i. Very low	i. High
	II. Single Mating	ii. High	ii. High	ii. Low
	III. Multiple mating	iii. Rare	iii. High	iii. Rare
9.	Fat body content category			
	I. High	i. Very low	i. Low	i. High
	II. Medium	ii. High	ii. Low	ii. Rare
	III. Low	iii. Moderate	iii. High	iii. Rare

Studies on Identification of Migratory Phases in Lepidopterous Pests

Extensive work was done at J.N.K.V.V. Jabalpur and B.H.U. Varanasi to study the migratory nature of few species of lepidopterous pests collected in light trap in high populations at the end of *Kharif* or *Rabi* season. Adults were examined every week to observe seasonal changes in various characters related to the identification of migratory phase of a population. Observations were made on sex ratio, state of ovary development, mating condition (unmated/mated) and fat body content (as energy reserve).

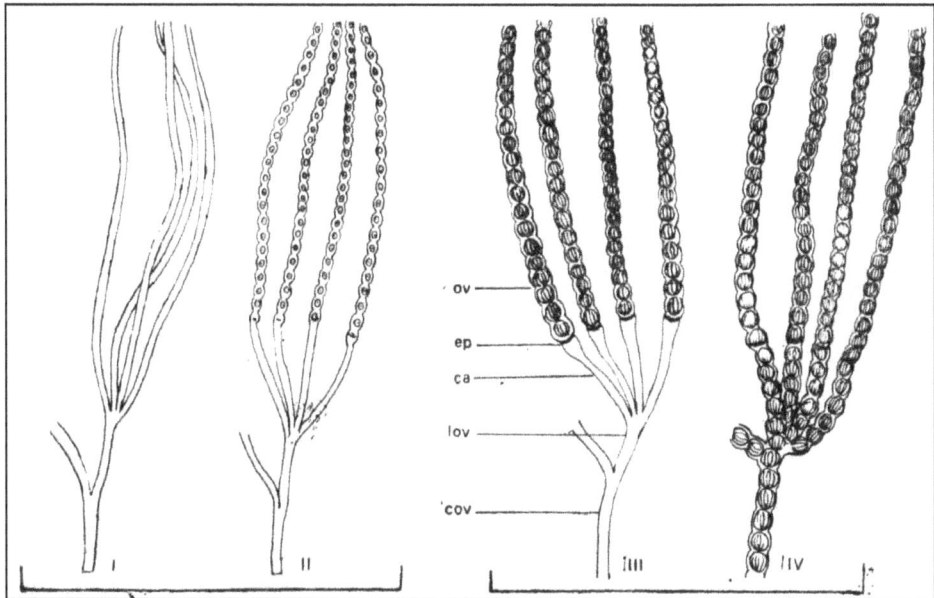

Figure 34: Different Stages of Ovary Development in Female Moths of
***Heliothis armigera* Collected on Light Trap.**

Ov: Ovarioles; ep: Epithelial plug; ca: Calyx; lov: Lateral oviduct; cov: Common oviduct; Grade I: No development; Grade II: Partial development; Grade III: Complete development but pre oviposition stage; Grade IV: Ovarioles fully developed.

As a sample case, various categories of ovary development and data on variation in mating frequency in females of *Heliothis armigera* are presented in Figure 34 and Tables 18 and 19, respectively.

Species-wise Observations are Summarized Below

1 (a) Observations on *Heliothis armigera* at Jabalpur (Vaishampayan and Verma, 1987).

Observations were made by analyzing the trap catches for a period of 3 years (1980-81 to 1982-83).

Table 18: Seasonal Variation in Ovary Development of *H. armigera* Moths collected on Light Trap at Jabalpur during the Active Seasons of 1980-83

Period	Percentage of Females in different Ovary Development			
	1980-81		1982-83	
Weather Week Number	Nil/Partial Development	Complete Development	Nil/Partial Development	Complete Development
Dec 49	62	38	75.0	25.0
50	50	50	10.0	0
51	–	–	29.0	71.0
52	33	67	0	100
Jan 1	10	87	0	100
2	7	93	0	100
3	Not dissected		–	–
4	9	91	0	100
Feb 5	12	88	14.6	85.4
6	67	33	27.3	72.7
7	0	100	56.3	43.7
8	5	95	62.9	37.1
9	–	–	56.7	43.3
Mar 10	60	40	61.7	38.3
11	80	20	93.5	6.5
12	37	63	90.3	9.7
13	50	50	89.7	10.3
Apr 14	67	33	60.9	39.1
15	84	16	65.6	34.4
16	–	–	66.7	33.3
17	–	–	100	0

Cat. I and II – Nil/Partial development Cat. III and IV – Complete development of ovarioles.

Table 19: Seasonal Variation in Mating Frequency of *H. armigera* Moths Collected on Light Trap during 1982-83 Active Seasons at Jabalpur

Period Monthly Interval	Percentage of Females with Various Spermatophore Number						
	0	1	2	3	4	5	6
Jan.	0.0	30.8	15.4	15.4	23.0	7.7	7.7
Feb.	38.9	12.5	20.1	12.5	8.3	5.5	2.1
Mar.	86.8	4.8	3.6	1.8	24	0.6	0.0
Apr.	67.7	16.9	5.6	2.8	4.2	1.4	1.4
May	78.3	18.1	3.6	0.0	0.0	0.0	0.0
June	88.2	8.9	2.9	0.0	0.0	0.0	0.0

Evidences based on the trend in adult activity and associated changes in sex ratio maturation of ovary and frequency of mating of females lead to following conclusions.

a. **December catch:** Trap catches during December are pre dominated by immigrants (incoming migrants) Ratio of 'fully developed' and 'No development' in ovaries was 50: 50.

b. **January and February catch:** Adult activity was moderate to slightly high. Proportion of fully developed ovaries (gravid females) and mated females (multiple mating) was highest. This is a characteristic of Resident Population (locally bred).

c. **March/April catch:**

    i. Adult activity was at highest level.

    ii. Sex ratio – Proportion of females was higher than males.

    iii. Ovary development was ceased. Ovary development was absolutely nil (thread like) in almost 100 per cent females.

    iv. Proportion of Unmated females was very high (68 to 88 per cent)

These are the characteristics of Emigrant Phase of a population (outward migration enmass). This explains the disappearance of pest in the region during summer and rainy season.

1 (b). Observations on *Heliothis armigera* at Varanasi (Vaishampayan S. Jr. and H.N. Singh, 1995)

Light trap catches were analyzed during 1991-92 to 1992-93 seasons to study the migratory nature of the pest.

    i. Evidences based on seasonal changes in the wing span, body length, fat body content, ovary condition and frequency of mating of females collected in the trap proved in a row, the migratory nature of the pest.

    ii. Distinct changes were observed in various characters of trap catches studied at a time when activity was highest in March and April. Most predominant changes were (i) cessation of ovary development, (ii) Un mated condition most prevalent and (iii) accumulation of fat bodies (higher fat body contents) All the evidences in a row established migratory nature of *Heliothis armigera* population was in Emigration phase in March and April indicating outward migration of adults enmass.

2. Observations on Teak Defoliator *Hyblaea puera* at Jabalpur (Vaishampayan, Verma and Nema, 1984)

    i. Light trap catches were analyzed every week during 1980 and 1983 active season to study the seasonal changes in sex ratio ovary development and mating frequency.

    ii. Principal activity was restricted to only 10 to 12 weeks during July and August. Pest was totally absent in the region for almost 9 months until next monsoon season.

iii. In Early July females were dominated by developed ovaries (about 100 per cent), mating frequency was higher and fat body content in medium category.

These are the features of Immigrant phase of a population.

iv. In August, females were dominated by - Nil or partial ovary development (in more than 75 per cent population), unmated females (in 65 to 86 per cent population) and high fat body content.

These are the features of Emigrant phase of a population.

These observations provided evidences on migratory nature of a species *i.e.* Immigration in July and Emigration in August. There was no Resident (locally bred) Population of adults.

3. Observations on *Plusia orichalcea* (Vaishampayan S.Jr. and H.N.Singh, 1996)

Trap catches collected during the period 1991-92 and 192-93 Rabi season were analyzed for various characters. Activity of pest was low to medium in January and February (monthly catch- 34 and 277 moths). Population suddenly raised to very high level in March (64,566 moths)

A. December to February Catch

i. Ovary Development- Gravid females with developed ovaries in various stages was prominent (75-100 per cent population).

ii. Sex ratio was lower (21-31 per cent) compared to later period.

iii. Mating frequency – (females with one or two mating was prominent)

iv. Fat Body content- females with low fat body content were higher (65-88 per cent).

These are the features of mixture of Immigrant (December) and local breeding/ Resident population

B. March Catch

i. Sudden rise in activity- very high population (64566 moths).

ii. Sex ratio- sudden rise to 50 per cent level.

iii. Ovary Development ceased- nil/initial stage was prominent (87-96 per cent)

iv. Mating frequency- unmated condition was very high (85-96 per cent)

v. Fat body content was very high (85-95 per cent)

These are the features of a migrant population in Emigration Phase.

4. Observations on *Plusia chalcytes* at Jabalpur (Mahobe and Vaishampayan 1990).

Light trap catches collected during 1983 and 1984 *Kharif* seasons were analyzed for various characters related to migration. On the basis of various characters observations were made on the condition of ovary development, mating frequency and fat bodies content, as observed in other species described above. Results indicated existence of migratory nature of the pest identifying three phases as below:

i. Immigrants during July and August,

ii. Locally bred 'Resident population' during October and

iii. Emigrants during November

5. A. Observations on Black Cutworm *Agrotis ipsilon* (in Rabi) at Jabalpur (B. Ennerjeet 1988), Major Advisor Vaishampayan S.M./Ph.D. Thesis (Unpublished)

Result of 3 years study (1984-85, 1985-86 and 1986-87) have provided sufficient evidences on the migratory nature of the species, identifying such populations into three distinct phases namely.

i. **Immigrants**: (September, October, November)- Low sex ratio/single mating more common, ovarioles in II and III developmental stages were predominant, fat body content was medium, neuro secretary cells (NSC) and *Carpora allata* (CA) were more active.

ii. **Residential**: (December, January February) moderate sex ratio, mating frequency was higher (unmated females very low to Nil), fat reserve was very low, ovaries were fully developed, NSC and CA were inactive (very low)

iii. **Emigrants**: (March and April) adult activity was at peak, sex ratio was high. Population was dominated (high proportion) by unmated (virgin females), ovaries immature or nil development with very huge fat reserves, (Fat body content- very high); NSC and CA were more active.

5. B. Observations on Black Cutworm *Agrotis ipsilon* at Varanasi (Vaishampayan S. Jr. and H.N. Singh, 1994).

Observations were made during 1991-92, 1992-93 and 1993-94 crop season.

i. Activity was restricted during December to May only.

ii. In January and February all stages of ovary development were found in females (80 to 100 per cent), fat body content was very low and ratio of mated females was very high(90 to 100 per cent).

These are the features of Resident Population.

iii. In March and April-Proportion of immature ovaries (grade I) was very high. Ratio of females with high fat body content was very high. Proportion of unmated females was also much higher.

These are the features of Outgoing migrants (Emigration phase).

11
Field Studies on Survey and Control of Economic Species

The material presented herewith in this chapter is a world review of field studies on various aspects of use of light trap related to survey and control of insect pest species including observations on their attraction to specific light sources or electric lamps actually used in the study, emitting radiation in a range varied from Ultra Violet (around 300nm), near ultra violet (360 to 380nm) to visible range up to 700nm.

The results of various investigations carried out as discussed above are reported herewith species wise in two sections.

Section A: Review of world literature

Section B: Review of Indian work on major species.

Review of World Literature on Major Pest Species

1. European Corn Borer Survey and Control

Kelsheimer (1935) conducted field and laboratory research under controlled conditions to determine the influence of different colours of lights on the behavior of the European corn borer moth. The light transmitted through all the filters was uniform and comparable. The moths responded in significantly greater numbers to the lights of short wavelength than to those of long wavelength. The blue light of the series attracted more moths than did the red light on the opposite end of the series.

In field tests Ficht and Anderson (1942) compared European corn borer moth catches using six lamps of each type. Results of field tests and laboratory studies in combine indicated that the greatest radiation attraction to the European corn borer moths was in the range of 320-500 nm. Taylor and Deay (1950) found that increasing the amount of near ultraviolet radiation increased the attraction of moths but not in direct proportion to the increase in radiant energy.

Corn Borer Control Activities with Light Traps

Ficht and Hienton (1941) found that infestations and populations could be greatly reduced but not eliminated by lighting cornfields with one 250-w. Mazda CX lamp and circular grid trap per acre. In 1939, reductions of corn borer infestation in a 10-acre cornfield by use of electric light traps averaged 75.3 percent below those in three adjacent fields. Also in 1940, reductions in a 12-acre lighted field averaged 66.7 percent below those in five adjacent fields. Adult corn borer catch per trap averaged 1,117 moths per season in 1939 and 3,218 in 1940. Deay and Foster (1944) reported that one trap with H-4 mercury vapor lamp per acre reduced the infestation of first generation corn borers from 27.5 to 14 percent in corn planted on June 1. One light trap per acre reduced the fall population of borers from 3.6 to 2.5 borers per plant in corn planted on June 1.

2. Codling Moth Survey and Control

Codling Moth Attraction to Light and Effects of Trap Design on Moth Capture

Recognition of the attraction of light to the codling moth more than a century ago is evidenced from a statement by Glover (1865)-"Bonfires in June evenings are recommended for the moth (codling)."

One of the earliest recorded experiments on the attractant value of various colours of light to the codling moth was made in New Mexico in 1915, 1916, and 1917 by Fite as reported by Eyer (1937). Nine incandescent lamps used were suspended over tubs of water which served to trap the moths. While only commercial types of tungsten filament lamps were used and the various colours incorporated in the globes presented equal emission of light, "the superiority of a blue or purple (violet) light over other colours was well demonstrated."

Laboratory investigations on the response of the Oriental fruit moth and codling moth to coloured lights were conducted by Peterson and Haeussler (1928) during the 3-year period 1925-27. They used clear incandescent lamps combined with several colour screens as attractants. Their studies showed that, if codling moths are given a choice of lights varying in colour from red to violet and the relative intensities of the coloured lights are approximately equal, practically all of the moths will go to blue and violet lights. Few or no adults are attracted by red light.

As reported by Marshall and Hienton (1935) studies were made during the period 1933 to 1934 in three sets *i.e.* (i) in apple orchards, (ii) in fruit packaging house (storage) and (iii) in a laboratory testing attraction of codling moths to a wide variety of lamps radiating energy in the visible and ultraviolet regions. Their early results indicated the most attractive region of the spectrum was between 300nm and 400nm, or near ultraviolet and violet. In the packinghouse darkened during the day and with an electrocuting UV light trap, operating both day and night, more than 98 percent of the emerging moths were attracted and killed (Davis, 1935). The maximum catch for a single day was 15,579 codling moths; and the total for the season, up to July 27 *i.e.* the end of the first brood emergence was 2,36,300 moths.

In field studies conducted during 1934 and 1935, comparing the relative attractiveness of blue, green, amber, red, and white incandescent lamps in electrocutor traps to the codling moth, Bourne (1936) found that codling moth showed a preference for the blue lamp.

Codling Moth Survey Activities with Light Traps

Hamilton and Steiner (1939) examined light trap captures at half-hour intervals on several different days during 1934 and 1935. They found that lamps did not begin to capture codling moths much before light receded to 0.2 ft.-c. or less. Although, moths came to traps all the hours from late dusk to early dawn, 85 percent entered the traps before 10: 30 p.m. and peak captures occurred between 7: 30 and 9: 30 p.m. Groves (1955) compared bait and light traps for catching codling moths in 1951 and 1952, using a Robinson trap (Figure 35) equipped with an 80-w. mercury vapor lamp. The single light trap caught considerably more codling moths than 12 bait traps in 1951 and 10 bait traps in 1952 with which it was compared.

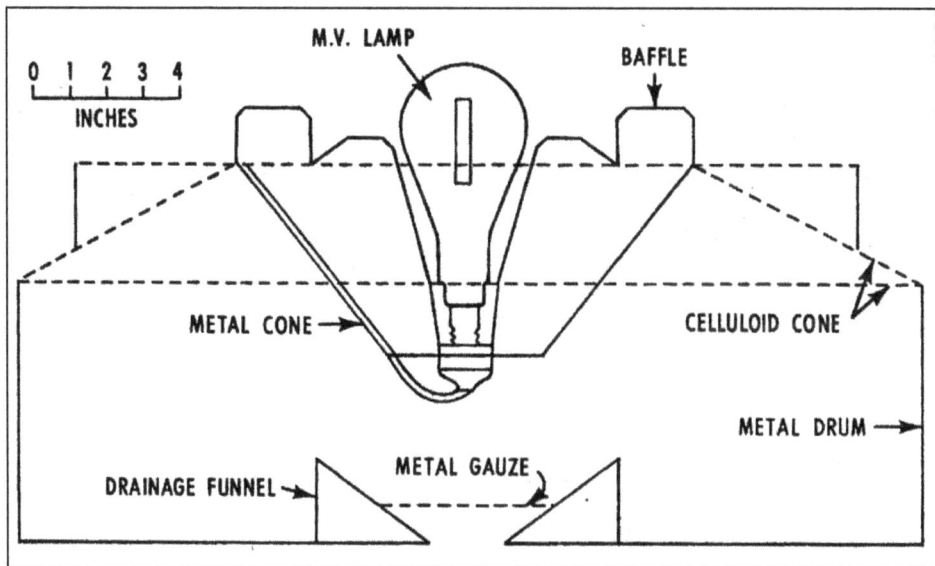

Figure 35: Cross Section of the Robinson Insect Trap Tested by Groves (1955) against Codling Moth using 80w MV Lamp.

Oatman and Brooks (1961) reported results of 5 years' experience with BL traps (1956-60), similar in design to the omnidirectional trap (see figure 25), and showed that it is an effective, additional survey tool for orchard insect populations. The BL trap proved to be several times more effective than the regular type incandescent light trap for surveying insect populations (Oatman, 1957). Its greatest value had been to time spray applications for the individual pests, especially the second generation codling moth.

In 1961, Madsen and Sandborn (1962) found that a funnel-type trap with a 15-w. BL lamp was very efficient in trapping codling moths. The codling moth was one of the species added to USDA's weekly Cooperative Economic Insect Report in 1963, when the number of insect species included in the list of those collected in light traps was increased from 9 to 20. Thus, there is international acceptance of the use of light traps as a determinant of codling moth activity.

Experiments on codling moth control by electric light traps were conducted in southern Indiana (1934-35) and in the Hudson Valley, N.Y., in 1936 by Hamilton and Steiner (1939). Codling moth infestation in the 5 ¼ acre light-trap area was reduced up to 44 percent compared to the surrounding check blocks. In 1935, seasonal conditions were extremely unfavorable for the codling moth, and the lighted area showed a 90-percent reduction in infestation.

3. Tobacco Hornworm and Tomato Hornworm Survey and Control

The attraction of adult tomato and tobacco hornworms to near ultraviolet radiation between 320 and 380 nm was reported by Deay and Taylor (1950). The attractant lamps used were germicidal, black light, and blue with maximum radiation at 253.7 nm, 365 nm, and 440 nm, respectively. Of the three types of lamps used in 1948, the 360 BL lamps were outstanding in attracting 92.6 percent of both the species of hornworm moths captured by traps in open fields.

As reported by Deay (1961), the omnidirectional trap used in 1955 was developed by Taylor and used during the 1956-59 seasons in Indiana experiments to protect tobacco from tobacco and tomato hornworms. This trap included a single, vertical, 15-w. BL lamp mounted in a single vertical baffle. The baffle was mounted virtually within a funnel, the baffle extending an inch above the funnel discharge opening to a height of 19 inches above the funnel rim. The funnel was 14 inches in diameter at the top, 2 inches in diameter at the bottom, and 12 inches high. The collection chamber was made of 9-inch-diameter metal furnace pipe, 24 inches long, and held against the funnel by spring tension. Bell (1955) studied 21 commercially available electric lamps, radiating energy in parts of the spectrum from the ultraviolet through the infrared as attractants for moths of the tomato and tobacco hornworm species. He concluded that five of the lamps, with high radiation outputs between 320 and 400 nm, were more attractive.

Menear (1961) found that the aggregate reaction of 531 individually tested moths was greatest at 315 nm when compared with responses at bands spaced 20 nm apart within the range of 315 to 455 nm. Further, he found that responses were nearly as good throughout the ultraviolet region as at 315 nm.

Experiments were conducted in six tobacco fields in Indiana during the 1956-59 seasons by Taylor and Deay (1961). Fourteen traps developed by Taylor were used in these tobacco fields to attract and capture hornworm moths.

Preliminary experiments in 1954 and 1955 indicated that a trap equipped with one 15-w. BL lamp would protect the tobacco within a radius of 100 to 120 feet from the lamp. Deay and Hartsock (I960) showed that, on an average, one trap equipped with one 15-w. BL lamp decreased the number of infested plants by 73.5 percent and the amount of leaves eaten by 77.4 percent in an area 100 feet from the trap. As a result, they designed and installed in 1962 a large-scale light trapping experiment in an area near Oxford, N.C. A circular area, 12-miles in diameter, was covered with about three light traps per square mile. These 324 traps were similar to the design which later was adopted as a standard for survey traps but with 18-inch funnel top width. The area trapped was increased to a 20-mile diameter circle in August 1964, with a total of about 1,100 light traps (Stanley and Taylor, 1965). Results of these experiments for the 1962-64 were reported by Lawson and others (1963 and 1966), Stanley and others (1964 and 1965), and Gentry and others (1967) while Lam and others (1968) reported the results for the years 1965-66. During 1962 to 1966, the estimated hornworm reductions in the trapped area ranged from 54 to 94 percent, depending on the species, sex, and year.

Overall results of these experiments suggested that severe suppression of tobacco hornworm populations in the United States does appear to be a possibility if black light traps are used in sufficient density and maintained for several years Cantelo and others (1972).

4. *Heliothis zea* (Corn Earworm, Cotton Bollworm and Tomato Fruit worm) Survey and Control

Pfrimmer (1955) conducted studies during 1954 to compare the responses of different orders of insects to three sources of BL radiation namely: a 15-w. BL lamp, a 15-w. BLB lamp; and three 2-w. argon glow lamps. The BLB trap caught twice as many insects as the BL trap and about 12 ½ times as many as the argon trap. Although the BLB lamp attracted nearly 2 ½ times as many Lepidoptera as the BL lamp, the bollworm response to the BL lamp was greater than to the BLB, and much greater than to the argon lamps.

Contrary to several reports of significant reduction in pest population using light traps there are few reports of almost total failure of light traps in controlling *Heliothis zea*. No significant I egg masses or larval population was found in trap area compared to non trap control area. Noble, Click, and Eitel (1956) evaluated attempts to control insects with light traps in certain cotton, corn, and vegetable crops in a large-scale experimental installation at Batesville, Texas. In 1955, Growers operated 142 light traps on five adjacent farms comprising a block of approximately 3,000 acres. This acreage, called the 'trap area' was compared with check fields outside of the area. The traps were 24- by 25-inch electric-grid type, equipped with two 15-w. BL lamps. Infestation records on the corn earworm were obtained from cornfields by Noble where no insecticides were used. Infested ears averaged 99.5 percent in the light-trap area and 99.3 percent in the check.

Light traps also appeared ineffective in controlling the bollworms in cotton. Results of a large-scale field evaluation of electric insect traps to reduce bollworm populations in Reeves County, Tex., during 1965 were reported by Sparks (1967). About 16,000 acres in a 12 by 35 mile belt devoted primarily to cotton were equipped with about 2,000 electric traps of four basic designs. All the traps were equipped with BL lamps as attractants.

In first experiment, comparisons were made among one untrapped and three trapped fields. Oviposition records, taken at irregular intervals in trapped and untrapped fields, failed to indicate that the trapping program consistently produced lower oviposition counts. In a second experiment Sparks compared bollworm oviposition and larval count records in a trapped and an untrapped field utilizing chemical control versus no chemical control. No insecticides were used on the trapped field. The egg count record indicates that the insect traps were as efficient as the 10 applications of insecticides in keeping the egg count under control. Again, the light traps appeared to be as effective as the 10 applications of insecticide in controlling populations of bollworm under the conditions of this experiment. In closing, Sparks commented: "the system of using traps with BL lamps to reduce insect populations is certainly not a cure for all the insect problems of cotton growers; neither is it something to be overlooked."

Graham and others (1971) conducted a study on corn earworm control in corn with a rather dense installation of BL traps in Guemez, Tamaulipas, Mexico, during 1966 and 1967. An installation of 79 suction-type traps, with one 15-w. BL lamp was made in an irrigated field of approximately 20 ha. The traps were placed at intervals of about 200 ft. through the cultivated area. They reported as follows: "data suggest that such an installation of light traps is not useful for protecting an individual field of corn from damage by the corn earworm.

5. Pink Bollworm Moth Attraction to Light and Effects of Trap Design on Moth Capture

Maxwell-Lefroy (1906) is the first entomologist, found to record capturing pink bollworm moths in lamp traps. Husain and others (1934) reported the attraction and capture of pink bollworm moths by light traps in the Punjab (India) during 1929-31, using an incandescent lamp of 200 cp. as the source of light. Moths were trapped in the field from the middle of July to the first week of November. The largest number collected during 3 years, however, was from the middle of September to the middle of October.

Glick and Hollingsworth (1955) conducted laboratory tests in 1953 with 28 lamps or combination of lamps having radiation outputs that covered various regions of the electro-magnetic spectrum between 184.9 nm (ozone lamp) and 1,200 nm (infrared drying lamp). Of the several sources tested, only one single lamp proved to be more effective than the 15-w. BL fluorescent lamp a 100-w. spot-type, mercury vapor lamp (H100-SP4) equipped with a filter which transmitted primarily in the near ultraviolet region. The principal radiation from this lamp is in the near ultraviolet region of the spectrum. Lamps that had their principal radiation in the visible portion of the spectrum attracted fewer moths.

Further studies on the attraction of pink bollworm moths made by Glick, Hollingsworth, and Eitel (1956) in 1954, verified the greater response to lamps that radiated in the near ultraviolet (black light) region. Laboratory investigations of the spectral response characteristics of pink bollworm moths were conducted by Hollingsworth in 1957, 1958, and 1959. The peak response was indicated at approximately 515 nm (green). Decreased response occurred in the vicinity of 415 nm and then a secondary peak response occurred in the near ultraviolet region at about 365 nm. There was very little response to wavelengths longer than 600 nm or shorter than 300 nm (Hollingsworth, 1961).

6. Cabbage Looper Moth Attraction to Light and Effects of Trap Design on Moth Capture

Pfrimmer (1957) studied the response of insects (including the cabbage looper) to different sources of black light. He used a 15-w. BL and a 15-w. BLB lamp in traps similar in design to that developed by Hollingsworth for insect surveys (Figure 32). The third trap with a 100-w. mercury vapor lamp was the same as the one he used in 1954. During both years, the BL lamp attracted the greatest number of moths, with the mercury vapor lamp second highest in 1955 and lowest in 1956.

Hollingsworth, and Hartstack (unpublished) conducted lab studies on the spectral response of cabbage looper moths in a V shaped test chamber during 1966. Results showed a peak response by the cabbage looper in the near ultraviolet region with a secondary peak in 475-575 nm regions.

Parencia and others (1962) collected cabbage looper moths at Waco, Texas, in a light trap for a period of 6-years during 1956-61. The light trap used in 1956 and 1957 was of the horizontal, gravity-unidirectional type with 100-w. mercury vapor lamp developed by Taylor for use in studies of European corn borer. The trap used during the other four years was the omnidirectional, gravity-type trap developed by Hollingsworth with 15-w. BL lamp mounted vertically. Annual cabbage looper moth catches were 49,863 in 1956 and 9,902 in 1957 by the mercury vapor lamp trap. Similar catches by the trap with BL lamp varied from 15,018 to 25,336 during 1958-61. Moths were usually collected beginning in April and ending in November, with maximum catches in August.

Extensive experiment on the combined use of sex pheromone and electric traps for cabbage looper control was initiated near Red Rock, Ariz., in March 1967, as reported by Wolf and others (1969). The site for this experiment was a 3,110-acre ranch of which 2,240 irrigated acres were cropped. Lettuce was grown on 1,800 acres-1,000 for a fall crop and 800 for a spring crop. Cotton was usually grown on 200 acres each year.

Experiment included 415 barrel mounted traps for insect control and 43 traps for monitoring adult moth populations. Two 15-w BL fluorescent lamps were used in the traps with 4 baffle plates. The catches of adult cabbage loopers were assembled into five groups based on the distance from the outside border of the trapped area. A summary of the numbers of moths caught per trap per night during the period of 3-years from 1967-69, is shown in Table 20. The reductions in adult looper catch from 1967 to 1968 and 1969 are clearly evident, with the exception of males at the border. The reduction from the border to the center of the trapped area is also evident when data from a 5-year average are considered.

Table 20: Summary of Light Trap Collections of Cabbage Looper Moths, 1967-69

Distance from Border (ft)	Cabbage Looper Moths Caught per Trap per Night					
	Female (Numbers)			Male (Numbers)		
	1967	1968	1969	1967	1968	1969
0 (Border)	9.16	3.97	5.99	13.95	10.60	13.47
1000	11.12	3.04	4.37	15.83	6.35	9.32
2000	12.49	2.44	3.23	15.79	5.86	5.45
3000	5.62	2.36	2.15	5.97	4.36	5.10
4000 (Center)	4.26	2.12	2.35	6.72	4.50	4.16

7. European Chafer Survey and Control

Adults of European chafer *Amphimallon majalis* (Razoumowsky) were found to be strongly attracted to radiation from fluorescent BL lamps. Tashiro and Tuttle (1959), experimented with omnidirectional gravity type light trap with four baffles using 15-w BL lamp. Traps with this lamp captured up to 70 times as many beetles as the most attractive chemically baited traps when exposed to very low populations late in the 1958 season. Light traps caught beetles on nights when none were seen in flight. Studies were conducted by Tashiro, Hartsock, and Rohwer (1969) and resulted in the development of the European chafer beetle survey trap, also described by Hollingsworth and others (1963).

The efficiency of BL traps as control tool was studied by Fiori in 1967 (unpublished). He operated gravity type light trap with 4 baffles on alternate nights using 30-w BL lamps in 1967 under a 20' tall Poplar tree. The data indicated that operating traps captured 80 to 100 percent of the beetles present.

8. Southern Potato Wireworm Survey and Control

Slingerland (1902) reported that too few Elaters or click beetles were taken in one trap lantern from May 20 to October 1, 1892, to be considered of economic importance. Bogush (1958) reported the attraction of 10 species of click beetles (Coleoptera Elateridae) to light traps in Middle Asia between 1930 and 1934. He used a light trap with a 500-w. electric lamp and listed catches of 20,000 to 30,000 Elaterid in a single night (1936).

Adult southern potato wireworms (*Conodenis falli* Lane) were caught during each month of the year in a light trap equipped with a 15-w. fluorescent BL lamp operated continuously between 1956 and 1967 in one location over sod near Charleston, S.C.(Day and Reid 1969). A survey light trap conforming to Entomological Society of America standards, shown in Figure 25, was used in these studies. The largest numbers were taken between June and September and the smallest between December and March. Catches in this trap during midsummer was highest between 8 and 9 p.m. Differences between catches in traps with the lamps positioned at ground level and at 2, 4, 6, 8, 10, and 18 feet above ground level were not significant. In limited tests averages of 18,728 and 4,357 adults were caught in the light traps during the oviposition season in 1965 and 1966, respectively.

An experiment on the population suppression of the southern potato wireworm through the use of gravity-type light traps was conducted during the period 1968 to 1970 (Day and Crosby, unpublished). A 16-acre field in an isolated area was surrounded by 18 survey light traps (conforming to E.S.A. standards,) each with a 15-w. BL attractant lamp at approximately 200-foot intervals around the edge of the field. In these trials total number of southern potato wireworm adults caught during the period April to September was 114,139 beetles in 1968 and 112,195 in 1969, and it was 62,954 in 1970.

9. Hickory Shuckworm, Pecan Nut Case Bearer and Pecan Leaf Case Bearer Survey and Control

Tedders and Osburn (*1966*) conducted experiments during 1964 at Albany, Ga., to determine whether an insect trap fitted with a BL lamp would attract and trap insects that attack pecans. An omnidirectional, gravity-type trap was used, having four vertical baffles that surround a 15-W BL lamp and mounted vertically over a funnel with a collecting can.

They found that three economically important pecan insects, the hickory shuckworm (*Laspeyresia caryam*, Fitch) the pecan nut case bearer (*Acrobasis nuxvorella* Nuenzig), and the pecan leaf case bearer (*Acrobasis juglandis* (Le Baron)) are highly attracted to BL lamps.

A 3-year study was made by Tedders, Hartsock, and Osburn (*1972*) in an 8-acre pecan orchard to determine whether the hickory shuckworm could be suppressed with a high density of BL traps. Thirty-three survey light traps with 15-W BL lamps (conforming to E.S.A. standards) were used in the orchard during 1967 and 1968. The percentage of shuck infestation inside the orchard ranged from 17.6 to 1.2 and from 74.8 to 35.9 outside the orchard. The percentage of shuck infestation within the orchard was 17.2 in 1967. 10.6 in 1968, and 1.2 in 1969. Suppression of the shuckworm with light traps was found to be comparable to suppression with recommended insecticide treatments, under the conditions tested.

10. Control of Insect Pest in Storage Godowns

i. For Grain Storage Pests

Stermer *et al.* (1959) made comprehensive laboratory studies of the spectral response characteristics of seven species of stored-product insects. Nine narrow wave length bands (approximately 20 m/i in width) of radiation at equal physical intensities were used for the tests. These wavebands included 280.4 m/i in the ultraviolet and 600 m/i in the orange region of the spectrum. Four of the species used, the almond moth, the Angoumois grain moth, the lesser grain borer, and the red flour beetle preferred a wave band which peaked near 500 m/i in the green portion of the spectrum. A secondary peak of response was noted in the region between 334 and 365 m/i. One species, the Indian-meal moth, showed a peak response to wave bands between 334 and 365 m/i with a secondary peak at approximately 500 m/i. The rice weevil showed no significant preference for the various wave bands and one species, the flat grain beetle, did not react in sufficient numbers to permit analysis. All species responded poorly to wave bands at 600 m/i. There was practically no response at 280.4 m/i.

As reported by Dalvania the UV light trap can be placed in food grain storage godowns at 1.5 m above ground level, preferably in places around warehouse corners, as it has been observed that the insect tends to move towards these places during the evening hours. The trap can be operated during the night hours. The light trap attracts stored product insects of paddy like lesser grain borer, *Rhyzopertha dominica*, red flour beetle, *Tribolium castaneum* and saw toothed beetle, *Oryzaephilus surnamensis* in large numbers. *Psocids* which are of great nuisance in godowns are also attracted in large numbers. Normally 2 numbers of UV light trap per 60 x 20 m (L x B) godown with 5 m height is suggested. The trap is ideal for use in godowns meant for long term storage of grains, whenever infested stocks arrive in godowns and during post fumigation periods to trap the resistant strains and left over insects to prevent build up of the pest populations. In godowns of frequent transactions the trap can be used for monitoring.

ii. For Apple Storage (Packaging House)

As reported by Marshall and Hienton (1935) studies were made during the period 1933 to 1934 on control of codling moths in fruit packaging house (storage of apples). In the packinghouse darkened during the day and with an electrocuting UV light trap, operating both day and night, more than 98 percent of the emerging moths were attracted and killed (Davis, 1935). The maximum catch for a single day was 15,579 codling moths; and the total for the season, up to July 27 *i.e.* the end of the first brood emergence was 2,36,300 moths.

Marshall and Hienton (1938) suggested to apple producers that one or more light traps in the orchard around the packinghouse, or other farm buildings where apples have been stored, is an excellent means by which the daily flight activities of this insect (the codling moth) may be determined as an aid in the timing of sprays, thinning the fruit and timing other orchard operations which are determined by moth flight."

Review of Indian Work on Major Pest Species

Verma and Vaishampayan (1983) reported for the first time the seasonal activity of 8 different species of major crop pest monitored by light trap catches throughout the year from June 1981 to May 1982. Data of monthly distribution of trap catches of these species are presented in Table 21 followed by Species wise description in brief about the general importance and major active period in a crop season.

1. Teak Defoliator, *Hyblaea puera* (Cramer) Hyblaeidae

It is a very serious defoliator pest of teak (*Tectnona granodis*) being active during July and August only. Peak period of activity was recorded during July and August with two major peaks in the populations suggesting completion of only one generation in this active period. From October to May the activity was absolute nil.

2. Gram Pod Borer, *Heliothis armigera* (Hubner) Noctuidae

It is a polyphagous species and a major pest of gram, arhar, sorghum, tomatoes and cotton throughout India. Pest was active throughout the year. December to April was a major active period with 5 peaks in population. Highest monthly catch of 10405 moths was observed in February followed by April catch of 7791 moths. Four peaks were observed during July to November in a descending order of catch size. Thus, about 9-10 generations were completed in a year. The highest activity was recorded during February.

Table 21: Monthly Distribution of Major Insect Pest Species Collected on Light Trap at Jabalpur during 1981-82

Month	Total Catch of a Month								
	Hyblaea puera	Heliothis armigera	Spodoptera litura	Agrotis ipsilon	Plusia chalcytes	Plusia orichalcea	Holotrichia serrata	Gryllus spp.
June 81*	20	34	1	0	7	0	889	6394
July	1061	133	265	0	1234	278	27	2983
August	730	170	514	4	4101	1050	0	3519
Sept.	12	65	913	19	2124	623	10	2847
Oct.	0	38	463	2	599	223	37	1184
Nov.	0	31	66	25	179	167	0	512
Dec.	0	2205	41	33	46	175	0	136
Jan. 82*	0	360	14	29	70	384	0	178
Feb.0	0	10405	126	96	158	1417	0	227
March	0	4049	296	142	282	2108	0	1420
April	0	7791	339	163	190	1022	0	1902
May 82	0	286	126	26	21	133	0	970

*Collection of 12 days only.

3. Tobacco Caterpillar, *Spodoptera litura* (Boisd) Noctuidae

It is a polyphagous pest and has been reported to do serious damage as foliage feeder in crops like groundnut, tomatoes, cabbage, cauliflower and many *kharif* pulses like moong, urid and soybeans.

Pest was active throughout the year except June. Major active period was July to October with four distinct peaks in population. Activity was low during November, December and January and moderate during February to May with four more peaks. Highest activity was recorded during September in *kharif* season and April in summer season with total monthly catch of 913 and 339 moths respectively.

4. Black Cutworm, *Agrotis ipsilon* (Hufnagle) Noctuidae

It is a sporadic pest and appears in endemic form in certain pockets. Extensive damage to young growing seedling has been reported in several rabi crops like gram, wheat, peas and potato. November to May was major active period with maximum catch of 163 moths obtained in April. Compared to other Lepidoptera the activity of this species was relatively low.

5. Green Semilooper, *Plusia chalcytes* (Fabr.) Noctuidae

It is a major pest of many *kharif* pulses. Pest was active throughout the year. Like Spodoptera litura, the major active period was in *kharif* from July to November and moderately active during February, March and April with 5 and 3 distinct peaks in populations during these seasons respectively. Highest catch was recorded in August.

6. Cabbage Semilooper, *Plusia orichalcea* (Fabr.) Noctuidae

It is a major leaf feeding pest of many *kharif* and rabi pulses and vegetables like cabbage and cauliflower. It was active throughout the year except June. The major active period was recorded during February, March and April as well as during August and September with highest trap catch of 2108 moths observed in March. About 9-10 distinct major peaks were observed in trap catches in a year.

7. White Grub Beetle, *Holotrichia serrata* (Scarabaeidae)

White grub is a sporadic pest of many agricultural crops and teak seedlings in forest nurseries severely damaging root systems of host plants during August and September in many parts of India. Adult activity starts with the onset of monsoon and is limited to only one month during June to July in rainy season only. An abrupt rise in trap catch was noticed in June end, following pre-monsoon showers, which was sharply declined in the following weeks. With another little peak in October first week the pest disappeared in rest of the season until next monsoon season.

8. Field Cricket, *Gryllus* sp. (Gryllidae)

Very little is known about the status of these insects as a pest of agricultural crops. Pruthi (1969) and Richards and Davies (1980) reported their damage to many cultivated crops as soil pests damaging roots. Recently Chavan and Rao (1981) From N. Bengal recorded damage by these crickets to tobacco in seedling stages.

Both the species were collected on light traps in sufficiently high numbers throughout the year. The major active period of Gryllotalpa was from August to April with highest activity during October, November and January. The major active period of *Gryllus* spp. is June to October and again in March, April and May. The peak of activity was recorded during June with a monthly catch of 6394 adults. This range of activity in both the species indicates completion of more than one generation in a year as against only one generation reported earlier by Richards and Davies (1980) and Pruthi (1969).

In Orthoptera, it appears that full moon light increases the activity of insects resulting in higher trap catches. Several of the peaks in catches were coincided with full moon period.

Observations on Important Species of Insect Pest Collected in Light Trap at Jabalpur

Systematic data were collected on trap catches of seven species of Lepidoptera and three species of Orthoptera operating the trap every night round the year using 125 W \160 W mercury vapour bulb. Five years data of species wise yearly collection for the period 1982-1987 are presented in Table 22.

Observations on Beneficial Parasitic and Predatory Species of Insects Collected on Light Trap at Jabalpur

Studies were conducted during 1983 crop seasons to know if light trap operation has any significant effect on the withdrawal of population of beneficial insect (parasitic and predatory species) from the area, which may have adverse effect on natural bio-control activity in the ecosystem. In all 21 predacious and 8 parasitic species were recorded to appear in significant numbers (Tables 23 and 24). Highest trap collection was recorded in families of Carabidae and Coccinellidae, among predacious Coleoptera and Reduvidae among the Hemiptera. In parasitic species Ichneumonids and Braconids of the order Hymenoptera were predominant with maximum activity observed in March and April when *Rabi* crops are harvested (Khan, 1983).

Observations were made on total weight of all parasitic and predaceous species of insects as listed in Tables 23 and 24 compared to the total weight (biomass) of all the remaining insects (pest and non pest both) collected in light trap every day. Data of percentage of weight of insects in two different groups was computed on monthly basis during January to September for comparison.

Overall results indicated that withdrawal of beneficial insects was significantly very low – below 1 per cent from January to April and below 1.6 per cent during May to August. *Campoletis chloridae* (Ichneumonidae) an important larval parasite of *Heliothis* was of rare occurrence only. Results thus, reveal that light trap operations will have no significant adverse effect on existing natural bio-control activity in a crop ecosystem.

Upadhyaya *et al.* (1999) conducted studies on common predatory insects collected on light at Jabalpur. Observations were taken examining the trap catches

Table 22: Yearly Distribution of Major Insect Pest of Pulses Collected on Light Trap at Jabalpur during 1982-83 to 1986-87

Name of Insect Pest Species	Total Annual Catches per Trap					Major Host Crops
	82-83	83-84	84-85	85-86	86-87	
LEPIDOPTERA						
Gram pod borer *Heliothis armigera*	2080	14192	2505	4683	16291	Cotton, sorghum, sunflower,tomato
Tobacco caterpillar *Spodoptera litura*	2630	2691	3120	2290	2046	Cotton, soybean, sunflower, tobacco, groundnut, castor crucifers
Green semilooper *Plusia chalcytes*	4406	4723	6637	8611	6720	Cotton, soybean, sunflower, cruciferous, vegetables.
Cabbage semilooper *Plusia orichalcea*	2682	6514	7172	4134	6682	Cotton, sunflower, crucifers, soybean, vegetables.
Black cutworm *Agrotis ipsilon*	912	1099	1201	1922	1610	Rabi crops (Wheat, gram, potatoes, crucifers, linseed
ORTHOPTERA						
Field crickets						
ii) *Gymnogryllus humeralis*	15754	26577	31030	12460	22884	Soil pest-damaging seeds and seedlings, roots (*Kharif* and *Rabi*)
Mole crickets						
Gryllotalpa grylliotalpa	2773	4464	2405	1268	3411	Soil pest-damaging seeds and seedlings, roots (*Kharif* and *Rabi*)
Grass hoppers						
i) *Trilophidia cristata*						
ii) *Gastrimargus transverses*	1226	852	2690	309	1666	Polyphagous (*Kharif* Crops)

Table 23: Monthly Distribution of Various Parasitic Species of Insects Collected on Light Trap during 1983 Season at Jabalpur

Sl.No.	Name of Species	Total Collection per Month								
		Jan	Feb	Mar	Apr	May	Jun	July	Aug	Sept
A.	**Order – Hymenoptera**									
	Family – Ichneumonidae									
1.	Enicospiluscapensis	05	17	261	222	40	20	112	184	34
2.	Netelia sp.	00	00	00	00	22	34	163	97	25
B.	**Family – Braconidae**									
3.	Phanerotoma sp.	00	05	02	34	72	41	00	00	00
4.	Zelomorpha sp.	00	00	00	00	20	17	153	195	03
5.	Rogas sp.	00	00	29	55	00	00	00	55	26
6.	Homolobus sp.	28	29	107	106	07	08	00	10	07
7.	Macrocentrus sp.	00	00	05	49	22	21	105	235	35
8.	Yellicones sp.	00	00	00	00	00	00	151	95	20

Table 24: Monthly Distribution of Various Predaceous Species of Insects Collected in Light Trap during 1983 Season at Jabalpur

Sl.No.	Name of Species	Total Collection per Month								
		Jan	Feb	Mar	Apr	May	Jun	July	Aug	Sept
A.	**ORDER – Coleoptera**									
	Family – Carabidae									
1.	Brachinus sexmeculatus	27	3	7	14	1	15	139	169	52
2.	Chlaenites circumastus	65	5	28	13	2	21	231	291	89
3.	Chlaenius sp.	00	0	00	00	18	13	132	56	5
4.	Crospedophorus elegans	00	0	00	00	23	16	154	70	4
5.	Diplocheila polita (Feb)	23	5	5	2	9	40	544	378	211
6.	Deserida lineola (Mac)	00	0	00	00	12	18	91	92	00
7.	Zuphium olens (Rossi)	00	0	00	00	12	12	100	112	27
B.	**Family – Cicindellidae**									
8.	Cicindela melancholica	00	1	43	15	9	257	1683	395	12
9.	Prothyma peoxima (Ch)	00	0	21	10	6	36	124	80	6
10.	Cicindela venosa (Kol)	00	0	00	00	11	49	146	86	4
11.	Cicindela cardoni (Fle)	00	0	00	00	00	22	155	87	6
C.	**Family – Coccinellidae**									
12.	Cooccinela septumpuntata	21	1	25	00	00	00	00	00	00
13.	Menochilus sexmeculatus	1	2	2	00	00	00	00	00	00
14.	Brumus suturalis (F)	2	0	0	2	00	00	00	00	00
15.	Adalia bipuncata (Linn.)	00	0	00	00	3	3	2	00	00

Contd...

Table 24–Contd...

Sl.No.	Name of Species	Jan	Feb	Mar	Apr	May	Jun	July	Aug	Sept
							Total Collection per Month			
A.	**ORDER – Hemiptera**									
	Family – Reduvidae									
1.	*Sirthenes* sp.	33	9	64	27	17	66	159	126	22
2.	*Ectomocorts cordiger*	19	3	17	23	35	78	97	127	18
B.	**Family – Pyrrhocoridae**									
3.	*Antilochus cocquebenti*	00	0	13	16	59	74	122	139	14
B.	**ORDER – Neuroptera**									
C.	**Family – Chrysopidae**									
1.	*Chrysoperla* sp.	00	2	57	83	28	122	29	45	8
	ORDER – Dictyoptera									
A.	**Family – Mantidae**									
1.	*Statilis maculate* (Thu)	00	0	1	1	00	3	16	15	12
B.	**Family – Libellulidae**									
2.	*Libellula* sp.	00	0	00	2	1	10	57	54	107

Table 25: Monthly Distribution of Trap Catches of Various Predatory Species

Sl.No.	Order/Family	Name of Species	Aug	Sept	Oct	Nov	Dec	Jan	Feb	Mar
1.	**Order Coleoptera**									
	Coccinellidae	Coccinella sp.	1508	2882	1715	141	0	0	0	0
		Adalia sp.	0	5	0	0	0	0	0	0
	Carabidae	Brachinus sp.	40	55	44	23	2	0	0	0
		Chalenites sp.	58	65	18	28	0	0	0	0
		Chaleius sp.	53	291	51	2	0	0	0	0
		Crospedophonus	50	35	35	14	0	0	0	0
		Diplocheila sp.	43	115	53	168	0	13	6	23
		Zuphium sp.	95	60	62	14	24	0	0	9
		Deserida sp.	65	51	25	102	4	0	0	4
	Cicindelidae	Cicindela sp.	15	23	12	0	0	0	0	0
2.	**Order Hemiptera**									
	Reduviidae	Sirthenea sp.	38	28	35	5	6	0	19	7
		Ectomocoris sp.	30	0	9	14	17	3	0	0
	Pyrrhocoridae	Antolochus sp.	15	21	67	0	0	0	0	0

3 to 4 times in a week. The record of average per day catch was maintained for computation of data on monthly basis. Data are presented in Table 25.

Analysis of Trap Catches Shows

i. Out of 13 species recorded only one species *i.e. Coccinella* sp. was most predominant with maximum monthly catch of 2882 observed in September. However, its activity was restricted to only four months August to November.

ii. During the period December to March the collection was very low, almost nil in almost all the species. Collection of *Coccinella* was absolute nil in all these months when activity of mustard aphid on crucifers is significantly high.

iii. *Campoletis chloridae*, a larval parasite of *Heliothis*, was found to be absent from trap catches while the activity of this parasite was very high in the field in gram crop. Around 20 to 40 per cent parasitization of early instar larvae of *Heliothis* was observed in field by this species.

iv. Overall results indicate that number of predatory sp. collected was relatively very low compared to huge collection of pest and non pest species observed in trap catches and hence, light trap can safely be used as IPM tool.

Comparative Efficacy of Various Light Sources against Lepidopterous Pest

The efficacy of various light sources attracting adults of *H. armigera, S. litura and A. ipsilon* was tested in the field at Jabalpur during 1977-78 (Vaishampayan and Verma, 1983). Considering the response of all the 3 species, mercury vapour 125 watt and UV 15 watt proved to be the most attractive light sources and the incandescent lamp 300 watt proved the least effective light sources (Table 26). Comparing response of 40 watt fluorescent lights, the blue colour, emitting radiation in the range of 450 to 480 nm wavelength proved to be most attractive to all the three species compared to green, yellow, red and white colours, although the total light intensity of light was highest in white. In general, shorter wavelengths elicited more response of moths than longer wavelength.

White Grub: Survey and Control

Observations on survey and control of white grubs as quoted by Dalvaniya D.K. (2010) from Deptt. of Entomology, C.P. College of Agriculture, S.D.A.U. India is summarized below:

White grubs are the cosmopolitan insect pests of agriculture, forest and pasture lands. The pest causes severe economic loss in upland paddy, finger millet, barnyard millet, maize, potato and many other vegetables, fruits and fodder crops in the hills of north-western Himalayan region, where nearly 40 species of this phytophagous pest were recorded.

Table 26: Relative Efficiency of Various Lights Sources against Major Species of Noctuids

Test No	Light Source Tested	Mean Catch/ Night	Relative Response	Test Period (Month)
A.	**Against Chickpea pod-borer (*Heliothis armigera*)**			
	Comparison with mercury vapour 125 Watts			
1.	M V 125 Watts v/s	25.33	1.0	Jan.
	Ultraviolet 15 Watts	3.00	0.12	Feb
2.	M V 125 Watts v/s	4.92	1.0	Feb
	Incandescent 150 Watts	0.30	0.06	
3.	MV125 Watts v/s	5.83	1.0	Feb
	Incandescent 300 Watts	1.16	0.20	
4.	M V 125 Watts v/s	2.28	1.0	Oct
	Petromax 500 candle power	1.14	0.5	
	Comparison with blue fluorescent tube light 40 Watts			
5.	Blue v/s	4.80	1.0	Mar.
	Green	3.400.44		
6.	Blue v/s	22.00	1.0	Apr.
	Yellow	3.72	0.17	May
7.	Blue v/s	12.22	1.0	Mar.
	Red	1.88	0.15	
8.	Blue v/s	8.33	1.0	Apr.
	White	4.77	0.57	
B.	**Against Tobacco caterpillar (*Spodoptera litura*)**			
	Comparison with blue fluorescent lamp 40 Watts			
9.	MV 125 Watts v/s	4.00	1.0	Dec.
	Ultraviolet 15 Watts	1.86	0.47	
10.	MV 125 Watts v/s	7.53	1.0	Oct.
	Petromax 500 candle power	2.92	0.39	
	Comparison with blue *fluorescent lamp* 40 Watts			
11.	Blue v/s	9.00	1.0	Mar
	Green	0.80	0.09	
12.	Blue v/s	4.33	1.0	May
	Yellow	0.77	0.19	
13.	Blue v/s	9.77	1.0	Mar.
	Red	1.00	0.1	
14.	Blue v/s	3.73	1.0	Apr.
	White	0.80	0.21	
C.	**Against Black cutworm (*Agrotis Ipsilon*)**			
	Comparison with mercury vapour lamp 125 Watts			
1.	Mercury vapour 125 Watts v/s	3.33	1.0	Jan.
	Ultraviolet 15 Watts	1.75	0.52	
2.	Mercury vapour 125 Watts v/s	2.50	1.0	Feb.
	Incandescent 150 Watts	0.00	0.0	
3.	Mercury vapour 125 Watts v/s	4.25	1.0	Feb.
	Incandescent 300 Watts	10.05	0.01	

Contd...

Table 26–*Contd...*

Test No	Light Source Tested	Mean Catch/ Night	Relative Response	Test Period (Month)
	Comparison with blue fluorescent lamp 40 watts (in different colours)			
7.	Blue v/s	3.50	1.0	Mar.
	Green	0.00	0.0	
8.	Blue V/s	7.66	1.0	Mar.
	Red	0.33	0.04	
9.	Blue v/s	3.83	1.0	Apr.
	White	0.83	0.22	

After Vaishampayan and Verma (1983).

The beetles emerge from soil from May to October. To combat the problem, Vivekananda Parvatiya Krishi Anusandhan Sansthan (VPKAS), Almora, ICAR designed a user-friendly low cost light trap for efficient mass trapping of beetles, to reduce the population of white grubs in soil. In all 61 light traps in different villages of Almora district on community basis were operated for 4 months during June to September 2006. During this periods total number of beetles trapped was 1,14,000 in Darima, 67,900 in Chausali. 46,600 in Tipola-Tunakot, 88,900 in Daulaghat-Govindpur, 76,100 in Manan and 31,400 in Bhagartola village.

Testing Response of White Grubs towards various Coloured Light Sources

This experiment was conducted in field during night from 9.00 to 9.30 hours in the area covering diversity of crops and forest vegetation. All the six light sources were arranged in line, 6 meter apart from each other to let the insects orientate toward their most favorite light colour suitably projected on white fabric screen. Data on relative response of white grub beetles, measured as percent attracted towards different coloured lights during night hours in the field are presented below:

Colour	Per cent Attracted
Red	2.20 per cent
Yellow	10.60 per cent
Green	4.70 per cent
White	18.00 per cent
Black (UV)	42.10 per cent
Blue	22.40 per cent
Total beetles caught	1020

Black light (UV) attracted the highest number of insects (42.1 per cent). Blue light was next attractant source (22.4 per cent) followed by white (18 per cent) in both the experiments conducted at different sites. The lowest number of insects were attracted towards red colour light (2.2 per cent) followed by green colour (4.7 per cent).

Management of White Grub Beetles by Light Trap (Vishwa Nath, 1983)

White grub, *Holotrichia consanguinea* Blanch; *H. serrata* Fabr; and *H. insularis* Brenske (Coleoptera: Melalonthidae) is a univoltine polyphagous and serious pest of variety of *kharif* crops in different states of India. The collections of beetles were made on the light trap periodically after emergence of beetles from 7.30 p.m. to early morning for a period of 19 days. Maximum numbers of beetles were attracted towards light trap between 8.30 p.m. to 10.30 p.m. and the peak period of beetle collection was 9.30 p.m. The maximum number of beetles was collected from neem (4974 adults) in the presence of light trap followed by guava (678 adults), ber (474 adults) and mango(8 adults). Light trap when kept alone also attracted 316 adults. The presence of trees like 'Neem' (*Azadiracta indica*) plays an important role in attracting beetles during night. Light trap can be kept near the vicinity of neem tree in the field so as to have maximum collection of the beetles.

Deployment of Novel Technologies for the Management of White Grubs in Lower Hills of New Himalayan Region on Community Basis (Rai and Sushil *et al.*, 2013)

White grubs, a group of destructive insect pests pf polyphagous nature, cause severe damage to crop plants in hill ecosystem The grubs with subterranean habitat feed expensively on the roots and adults defoliate the plants. Among different agroecological regions of India, the North-West Himalayan region comprising of the states of Uttarakhand, Himachal Pradesh and Jammu Kashmir have been identified as hotspots for white grub diversity. A two pronged strategies involving an efficient, light weight light trap for capturing the adults (beetles) and an entomo-pathogen *Bacillus Cereus* for the management of grubs were used. Large scale deployments of the above technologies were done on community basis at 5 locations including 4 villages and a farm of KVK, Uttarakashi district (Uttarakhand). Observation data are summarized in Table 27.

Table 27: Management of White Grubs through Insect Traps on Community Basis in Low Hills (<1 000 m amsl)

Adopted Villages	No. of Light Traps Installed	Beetles Trapped (in lakhs)			Beetle Reduction Over 2006 (per cent)	
		2006	2007	2008	2007	2008
Mohargaon	12	0.25	0.15	0.064	42.36	74.79
Barethi	10	0.22	0.15	0.069	34.39	69.29
Tuliyara	10	0.24	0.14	0.039	39.99	84.67
Gailari	9	0.26	0.14	0.038	44.99	83.98
Chinyalisaur	6	0.28	0.14	0.095	48.90	66.24
Total	**47**			**Mean**	**42.12**	**75.79**

Note: The trap is specific to the scarabaeids and was found to trap negligible number of beneficial insects.

Results of three year's experimentation revealed drastic reduction in beetel population to the tune of 75.8 per cent in low altitude villages. Reduction in grub population was also very significant, ranging from 74.11 per cent to 85.17 per cent in three years in different villages. Due to control of beetles and grubs population significant increase in percent yield of different crops was recorded, which varied from 39.0 per cent to 58.2 per cent in different villages. Sushil *et al.* (2008) have described in detail the species complex of white grubs of Uttarakhand Hills and their eco-friendly management using light traps.

Pest of Groundnut and Sorghum

The relationship between weather factors and light-trap catches of groundnut leafminer (*Aproaerema modicella*) (Lepidoptera: Gelechiidae) on groundnut (*Arachis hypogaea*) was studied in 1987-89 at Vridhachalam, Tamil Nadu, India (Baskaran and Mahadevan, 1994). The number of gelechiids attracted to a light-trap peaked in the 2nd fortnight of August and September in 1987 (11 peak catches), in the 1st fortnight of October in 1989 (7 catches) and in March and the 1st fortnight of April 1988 (6 catches). Results of the correlation studies indicated that rainfall alone exerted a significant positive influence, and an increase of 1 mm rainfall would increase the leafminer adults by 8.9.

A light trap study on *Chilo partellus* carried out in Karnataka, India, in 1987-88 and 1988-89 showed 2 peaks in activity periods in September to November, which coincided with the *kharif* sorghum crop; and January to April, coinciding with the rabi summer crop (Nandihalli, *et al.*, 1989). Tiwari, *et al.* (2001) developed a newly improved light trap designed for the survey detection and management of insect pests population of sugarcane based agro-eco- system at IISR (ICAR), Lucknow.

Work Done on Paddy Pest at AICRP (ICAR) Centres

Observations made under All India Coordinated Research Project (AICRP), ICAR on paddy (Anonymous, 1996). Light traps were installed in paddy fields and operated everyday in various paddy growing areas of different states of India. Observations were recorded regularly on major insect pest species collected in light trap. Data of highest weekly collection of four major species along with peak period of their activity, recorded during the year 1996 are presented in Tables 28 and 29.

Work Done on Various Paddy Pests at other Centres

Work was done at IRRI (Philippines) on tungro virus and vectors (Tiongeo, Hibino and Ling 1985). Tungro virus is the most important virus disease of rice in South and South East Asia and threatens serious reduction of rice production. Deptt. of IRRI conducted experiments to study the relationship between vector insects (leaf hoppers), tungro vectors (*Nephotetix* sp.) and grassy stunt vectors BPH (*Nilaparvata lugens*). Abstract data of 11 years (from August 1972 to December 1982) are given in Table 30.

Table 28: Observations on Weekly Light Trap Catches at Peak of Active Period Recorded at Various Centers of AICRP on Paddy during 1996

Name of Species: 1. Yellow stem borer, (*Scirpophaga incertulas*); 2. Rice gall midge, (*Orseolia oryzae*)

Sl.No.	Location	State	Weekly Catch at Peak of Active Period			
			Yellow Stem Borer		Gall Midge	
			Catch	Period	Catch	Period
1.	Rajendra Nagar	A.P.	1246	Oct.II	3750	Oct.II
2.	Maruturu	A.P.	5840 1416	May I Oct.II	– 22	– Oct.II
3.	Warangal	A.P.	669	Oct.II	401	Oct.II
4.	Rangolu	A.P.	628 603	Oct.IV Nov. IV	4172 –	Oct.IV –
5.	Aduthural	T.N.	491	Dec. IV	–	–
6.	Coimbatore	T.N.	76	Dec. IV	–	–
7.	Pattambi	Kerala	1007	Dec. III	518	Sept. IV
8.	Moncompu	Kerala	1240	Sept III	943 1609	Sept. III Dec. IV
9.	Mandya	Karnataka	1273 5754	July II Nov.IV	91	Nov. III
10.	KarjatSakoli	Maha.	–113	–Oct.I	117-	Sep. III-
11.	Raipur	C.G.	114	Oct.II	156	Oct. IV
12.	Chinsurah	W.B.	2283 2270	Oct.I Oct.II	538	Nov. I

Table 29: Period of Peak of Weekly Light Trap Catches Recorded at Various Centers of AICRP on Paddy during 1996

Name of Species: 1. Brown Plant Hopper (*Nilaparvata lugens*); 2. White Backed Plant Hopper (*Sogatella furcifera*)

Sl.No.	Location	State	Weekly Catch at Peak of Active Period			
			Brown Plant Hopper		White Backed Plant Hopper	
			Catch	Period	Catch	Period
1.	Rajendra Nagar	A.P.	20000 20950 21300	Sept.I Oct. II Oct. III	–	–
2.	Maruturu	A.P.	11222	Oct. IV	14100	Oct. II
3.	Warangal	A.P.	298	Oct. II	183	Oct. II
4.	Rangolu	A.P.	542	Oct. IV	177	Oct. IV
5.	Aduthural	T.N.	4055	Sept.II	–	–
6.	Coimbatore	T.N.	90	Dec. IV	76	Dec.IV
7.	Puducherri	–	128	Aug. IV	15	Aug.III
8.	Moncompu	Kerala	456	Dec. IV	–	–
9.	Pattambi	Kerala	–	–	280	Dec.II
10.	Mandya	Karnataka	144	Nov. IV	–	–
11.	Karjat	Maha.	2124	Aug. IV	–	–
12.	Chinsurah	W.B.	10379	Oct. I	4774	Oct. II
13.	Nawagam	Gujarat	–	–	306	Sept.IV

Ref. Annual workshop of AICRP on Paddy pest (1995).

Table 30: Number of *Nilaparvata lugens* Collected by Light Trap, Tested for the Grassy Stunt Disease Incidence at IRRI Farm 1972-82

Year	Number of Nilaparvata lugens		Grassy Stunt Infectivity	
	Collected*	Tested	Insect (per cent)	Incidence
1972	73,466	1087	1.83	High
1973	4,21,740	2636	2.86	Very high
1974	61,781	1900	0.93	Moderate
1975	8,823	2110	0.0	Low
1976	34,947	1790	0.0	Trace
1977	40,240	2116	0.72	Trace
1978	14,569	1394	0.35	Trace
1979	23,364	1214	0.0	Trace
1980	50,814	2597	0.23	Low
1981	3,96,238	2270	0.05	Low
1982	5,670	2134	0.34	Low

* From August to December collections.

Anuj,Bhatnagar *et al.* (1999) studied the effect of climate on the population buildup of rice insect pests using light traps in Jagdalpur, India, over a period of 4 years (1994-97). Yellow stem borer (*Scirpophaga incertulas*), leaf folder (*Cnaphalocrocis medinalis*) and caseworm (*Nymphula depunctalis [Parapoynx stagnalis]*) maintained a low level during the first 3 months of the cropping season. Relatively higher moth numbers were trapped during October, with highest activity in the final week.

Harinkhere, *et al.* (1998) studied the seasonal abundance of *Cnaphalocrocis medinalis* during the wet season for 8 years (1982, 1983, 1984, 1986, 1987, 1988, 1990 and 1991) at Waraseoni, Madhya Pradesh. Moth catches and field incidence started in the first week of August and showed a significant increase in the month of September. The peak period and field incidence of moths were observed in September followed by October. A significant positive association between light trap catches and field incidence reflected the severity pattern of leaf folders in the field. Pest incidence during a period of 8 years showed a regular incidence of leaf folder with considerable damage in the eastern part of Madhya Pradesh.

Kerketta, *et al.* (1990) reported the differences between the populations of *Sogatella furcifera* and *Nilaparvata lugens* caught in a light trap and counted in the field in 1986-87 in Madhya Pradesh, India, are discussed. The peak in the population of *S. furcifera* as measured by the light trap and by field collections coincided and the field population was correlated with the numbers caught in the light trap. Both pests were present from September to December.

The sex ratio and reproductive status of the rice pests *Scirpophaga incertulas* and *Cnaphalocrocis medinalis* captured in an ultraviolet-light trap in Tamil Nadu, India, were examined. More females than males of *S. incertulas* were captured and

58.6 per cent of females were gravid. The sex ratio of *C. medinalis* was more or less equal, and 61 per cent of females were gravid (Mohanraj *et al.*, 1989).

Nandihalli, *et al.* (1990) carried out light-trap studies during 1987-89 with *Scirpophaga incertulas* in rice fields in Karnataka, India. The result showed that the 1st peak of activity of the pyralid occurred during the period October-December and the 2nd from March to the end of May.

Pandey, *et al.* (2001) studied the effect of weather factors on the population build up of rice insect pests through light trap collection at Crop Research Station, Masodha, Faizabad, and Uttar Pradesh, India for over 10 years (1985-1994). Results revealed that yellow stem borer (*Scirpophaga incertulas*) maintained a low key during July and November of cropping season. Relatively higher moth numbers were trapped during August exhibiting their peak activity in the last week.

Regression models have been developed by Prasad *et al.* (2003) to predict the peak light trap catches of *Scirpophaga incertulas* (YSB) based on weather parameters such as maximum temperature, minimum temperature, morning relative humidity, evening relative humidity, sunshine hours and rain as independent variables. The weekly data obtained from the progress reports of the Directorate of Rice Research from 1975 to 2000 in Rajendra Nagar, Andhra Pradesh, India, were utilized to develop the models. The average peak height of LT catches was significantly lower during the rabi (840 per week) compared to the *kharif* season (1746 per week).

Light traps were used to monitor populations of the rice gundhi bug *Leptocorisa varicornis* (Fab.) at 2 sites in Uttar Pradesh, India, in 1980-81. At both sites, there were peaks in the populations in the 37th and 40th weeks of each year, when the rice was at the milky grain stage and when rainfall was relatively limited (Rai, *et al.*, 1990).

In a 26-year study on light trap catches of the green leafhopper (GLH), *Nephotettix virescens* made by Rai and Khan (2002) revealed that there was one peak catch during first fortnight of August in the *kharif* (main) season of rice crop in Patna, Bihar, India. In a study conducted from 1970 to 1995, the peak occurrence of yellow stem borer (YSB; *Scirpophaga incertulas*) on rice (*Oryza sativa*) was observed during the first fortnight of October (Rai, *et al.*, 2002). The period of peak occurrence of YSB coincided with the ranges of favorable environmental factors during the *kharif* seasons.

Observations made by Shrivastava and Mathur (1985) in rice fields at Raipur (M.P.) showed that the field populations and light-trap catches of *Nephotettix* spp. followed the same trend, the cicadellids appearing on the trap earlier than on the surrounding rice crop and also after the crop was harvested; this is due to the low tolerance of *Nephotettix* spp. to crowding and to the consequent need for dispersal flights, during which they are attracted to the light trap. Light-trap data can therefore be used to forecast *cicadellid* behavior and population peaks in the field.

Observations were made to monitor activity of plant hoppers in Karnataka, using light trap, sweep nets and dead hearts during the period 1985 to 1986 (Srinivasa, *et al.*, 1991). *Nephotettix* spp. and *Nilaparvata lugens* were present throughout this period but showed peaks of abundance in November and May.

Scirpophaga incertulas was present throughout the period, with low incidence in March, and had peaks in November and June.

Results of studies carried out during 1981-83 at 3 different localities in Karnataka, India, showed that the effect of moonlight was more pronounced than the effect of weather factors on light trap catches of rice pests (Srinivasa, *et al.*, 1990). Both in Bangalore and Raichur, *Nephotettix* spp. were more active around full moon than around new moon. Catches of *Nilaparvata lugens* and *Scirpophaga incertulas* in Bangalore were also greatest around full moon.

The effect of lunar cycle on light trap catches of the pyralid *Scirpophaga incertulas*, a pest (Stem borer) of rice, was examined during 15 lunar periods at Navsari, Gujarat, India. The highest catches were obtained during the new moon periods while full moon light interfered with trap efficiency (Pandya *et al.*, 1989).

Srivastava, *et al.* (1992) tested the ability of light traps to control pests in rice fields at 3 sites in Karnataka in the *kharif* seasons of 1989-91. The main pests caught in the traps were *Scirpophaga incertulas, Nephotettix* sp., *Nilaparvata lugens, Sogatella furcifera* and Pachydiplosis *oryzae* (*Orseolia oryzae*). The light traps caught sufficient insects to hold the populations below their economic thresholds in many cases and at three sites the pesticide use was reduced by 60 per cent, 50 per cent and 57.63 per cent.

Work on Forest Pests

Meshram *et al.* (1990) studied the seasonal activity of several pest species attacking teak and other forest plants grown in the forest nursery. A Pennsylvanian light trap unit equipped with a 160 W mercury vapour lamp was installed in the State Forest Research Institute Nursery at Jabalpur, Madhya Pradesh, during the 1984-86 seasons. The trap was operated every night from sunset to sunrise and insects collected every morning. Data are tabulated on monthly catches for 10 insect species, and the results for each of these are discussed individually. The pests were: the teak defoliator *Hyblaea puera*; the teak skeletonizer, *Eutectona machaeralis*; the white root grub, *Holotrichia serrata* (pests of teak *Tectona grandis*) pest); the Ailanthus web worm, *Atteva fabriciella*, a serious pest of *Ailanthus excelsa*; the polyphagous pest *Heliothis* (*Helicoverpa*) *armigera*; the cutworm, *Agrotis ipsilon*, an important nursery pest; the grasshopper *Hieroglyphus banian*, a pest of germinating seedlings of various forest trees; the mole cricket, *Gryllotalpa africana*, mainly a pest of agricultural crops; the field cricket, *Gryllus* sp., a nursery pest of various forest tree species; and the red cotton bug, *Dysdercus cingulatus*, a pest of cotton, *Abelmoschus* and other Malvaceae, and various forest trees in the nursery.

Seasonal activity of adults of teak defoliator *Hyblaea puera* and teak skeletonizer *Pyrausta machaeralis* (Pyralidae: Lepidoptera) was monitored by Light trap catches during 1978 and 1979 seasons at Jabalpur. Two years data of light trap catches of two species are presented in Table 31. Major active period of *H. puera* was July and August. September to June was off period with no indication of breeding in the area. Results indicated migration of pest from one area to other. Major active period of *P. machaeralis* was August to October. From December to June the activity was almost nil indicating no breeding of the pest in this period (Vaishampayan and Bahadur, 1981).

Table 31: Monthly Distribution of Light Trap Catches of Teak Defoliator, *Hyblaea puera* and Skeletonizer, *Pyrausta machaeralis* at Jabalpur

Month	Total Monthly Catch per Trap (Year-wise)			
	Hyblaea puera		*Pyrausta machaeralis*	
	1978	*1979*	*1978*	*1979*
Jan. to June	0	0	0	0
July	1610	8734	425	7184
August	149	5755	673	48655
September	2	228	976	45238
October	0	10	9484	60
November	0	4	1005	37
December	0	0	4	0

Sudden appearance of *Hyblaea* moths on light trap in very high numbers in the beginning of the season itself (early July), after an absolute absence of the pest in the preceding 9 months period, suggests the possibility of inward migration of moths (immigration) from other areas in the region. Similar is the case with *P. machaeralis*.

Use of Light Trap in Fish Culture

Heidenger (1971) described in brief the role of light trap (with UV lamp) in artificial fish culture. The terrestrial and aquatic insects collected in light trap and released in pond water serves as an artificial food to the fishes reared in the ponds. The experiments were conducted on effect of these insects on the growth of blue gill sunfish *Lepomis macrochirus* R. reared in the ponds. The bluegill sunfish was chosen for the experiment because aquatic and terrestrial insects normally constitute a large percentage of its diet. His observations are summarized below.

A pond's carrying capacity for fish is usually limited by its food supply. The addition of inorganic or organic fertilizers increases the production of natural forage organisms in a pond and the natural food supply can be supplemented with artificial food or by introducing forage organisms produced elsewhere. Ultraviolet light offers a means of supplementing the natural food supply with aerial insects derived from both the aquatic and terrestrial biotopes. Of the aquatic larvae that metamorphose and leave the water, 75 percent never return (Vallentyne,1952). Thus, considerable biomass which is potential forage is lost from the pond habitat. It is possible to recycle a portion of these adult insects into the pond by the use of ultraviolet light. Ultraviolet light can also be used to concentrate and make terrestrial insects available as food for fish. This is particularly significant in small lakes, since the terrestrial area is greater and can contribute a greater insect biomass.

Fifteen-watt fluorescent black bulbs were used as an ultraviolet light source throughout this study. These bulbs produce ultraviolet light (1800 fluorens) at wavelengths of 280 to 380nm and visible light (154 lumens) at wavelengths of 380 to 760nm. Peak radiation is at 350nm (Pfrimmer, 1955). In order to induce bluegill sunfish to feed at night, a 15-watt Pennsylvania light trap (Frost, 1957) was modified by attaching a 25-watt incandescent bulb below the ultraviolet bulb. The Pennsylvania light trap is characterized by four baffles mounted at right angles to each other. As insects spiral into the light source, many hit the baffles and fall into the water. When a light trap was used to attract insects to a cage containing fish, the funnel was removed, and the trap was suspended vertically over the center of the cage 15 centimeters above the water surface. This system allowed insects hitting the trap to fall directly into the water.

An experiment was designed to determine to what extent bluegill of the size used in this study were feeding on net zooplankton. Results show that by dry weight, 74 percent of the forage organisms eaten by bluegill were aerial insects, 14 percent were aquatic insects, and 11 percent were miscellaneous food items. Light traps increase the carrying capacity of ponds by adding aerial insects to the fishes' food supply. They also increase the vulnerability of aquatic organisms (Fore, 1969). Thus, fish expend less energy searching for and catching food organisms and more of the food intake can be utilized for growth. Two-fifths of the fish growth was attributed to food organisms entering the cage through the water, and three-fifths to aerial insects. Bluegill held in an illuminated cage at Fountain Bluff Pond consumed, in descending order of magnitude (by weight), the following: aerial insects → aquatic insects including larvae and → miscellaneous food items.

By dusk many insects were already attracted to the ultraviolet lights; therefore, before the carrying capacity is obtained, it would be possible to utilize a light trap for only 1 or 2 hours after dusk. This procedure would reduce the electrical cost 80 to 90 percent.

References

Agee, H.R., Webb J.C. and H.M. Taft. 1972. Activity of bollworm moths influenced by full moon. *Environ. Entomology.* 1 (3): 384-5.

Aino, S. 1954. Research on the phototropism of insects. (In Japanese, English summary): Nogyo-Kairyo- Gijitau-Shiryo, No. 52 *Min. Agr. and Forest, Japan.* 64 P.(See Iahikura 1967)

Allniazee, M.T. and Stafford, E.M. 1972. Seasonal flight pattern of the omnivorous leaf roller and grape leaf folder in central California Vineyards as determined by black light trap. *Environ, Entomol.* 1 (1): 65-8.

Anonymous 1996. Annual report of AICRP on paddy (ICAR).

Anuj Bhatnagar; Saxena, R.R.; and A. Bhatnagar, 1999. Environmental correlates of population buildup of rice insect pests through light trap catches. *Oryza.,* 36: 241-245.

Apple, J. W. 1957. Use of black light insect trap for detection of corn insects. *Proc. No. Cent. Br. Ent. Soc. Amer.* 12: 53.

Apple, J. W. 1962. Late-season corn borer and corn earworm moth flights in relation to larval population in sweet corn. *Proc. No. Cent. Br. Ent. Soc. Amer.* 17: 127-130.

Ark, H.V. and Pretorius, L.M. 1971. Sub sampling of large trap catches of insects. *Phytophytactica.* 3 (1): 29-32

Atwal, A.S., J.P. Chaudhary and M. Ramzan. 1969. Studies on the seasonal abundance of insects on light trap at Ludhiana. *J., Res. Punjab Agric. Univ.* 6: 186-96.

Ayyar,T.V.R. and Anantanarayanan, K.P. 1934. Insect Phototropism and its economic importance in *India. Madras Agric.* 22 (8): 268-73.

Ayyar,T.V.R. and Anantanarayanan, K.P. 1937. The stem borer pest of the rice (*Schoenobius incertulas* Wlk.) in South India. *Agric. and Livestock in India.* 7 (2): 171-179.

Baggiolini, M. and Stahl, Jr. 1965. Description of a model light trap for capturing insects. *Bull. de la Soc. Ent. Suisse* 37 (3): 181-190.

Ballou, H. A. 1920. Attraction of the pink bollworm to light. *In* Cotton and the Pink Bollworm in Egypt. *West Indies Imperial Dept. Agr. Jour.* 17: 276-279.

Banerjee, A. C. 1967. Flight activity of the sexes of crambid moths as indicated by light trap catches. *J. Econ. Ent.* 60 (2): 383-390.

Banerjee, S.N. 1985. Development of traps and insect trapping. Pp: 1-2. In: Use of traps for pest/vector research and control. Edt.: Mukhopadhyay, S. and Ghosh, M.R., Bidhan Chandra Krishi Vishwa Vidyalaya, Kalyani, W. Bengal.

Banerjee, S.N. and Basu, A.C. 1957. The Chinsura light trap. *Proc. Zool. Soc. 9*(1): 27-33. (State Agric. Res. Inst. Calcutta)

Bariola, L.A., Cowan, Jr. C.B., D.E. Hendricks and J.C. Keller. 1971. Efficiency of hexalure and light trap in attracting pink bollworm moths. *J. Econ.Entomol. 64*(1): 323-24.

Barnes, M. M., Wargo, M. J., and R.L. Baldwin,1965. New low-intensity ultraviolet light trap for detection of codling moth activity. *Calif. Agr.* 19(10): 6-7.

Barnes, M.M., Wargo, M.J. and R.K. Wagner. 1969. Reduction of treatment for codling moth in Chilean apple orchards by indexing with portable light traps. *J. Econ. Entomol* 62(3): 733-4.

Barrett, J. R., Jr., Deay, H. O., and J. G. Hartsock, 1971. Striped and spotted cucumber beetle response to electric light traps. *J. Econ. Ent.* 64 (2): 413-416.

Barrett, J.R. Jr., Deay H.O. and J.G. Hartsock. 1971. Reduction in insect damage to cucumbers, tomatoes and sweet corn through use of electric light traps, *J. Econ. Entomol. 64*(5): 1241-49

Barrett, J.R. Jr., Harwood F.W. and H.O. Deay. 1972. Functional association of light trap catches to emission of black light fluorescent lamps. *Environ. Entomol. 1* (3): 285-290.

Barrióla, L. A., Cowan, C. B., Jr., Hendricks, D. E., and J. C. Keller, 1971. Efficacy of hexalure and light traps in attracting pink bollworm moths. *Jour. Econ. Ent.* 64(1): 323-324.

Baskaran, R.K.M. and Mahadevan, N.R. 1994. Influence of weather factors on light-trap catches of groundnut leafminer (*Aproaerema modicella*) (Lepidoptera: Gelechiidae). *Indian J. of Agricultural Sci.* 64: 882-885.

Beall, G. 1938. Analysis of fluctuation in the activity of insects. A study on the European corn borer (*Pyrausta nubilalis* Hb.). *Canad. J. Res. 16*(3): 39-71.

Beaty, H. H., Luly, J. H., and D. S. Calderwood, 1951. Use of radiant energy for corn borer control. *Agr. Engin.* 32 (8): 421-422, 426,429.

Beck, E.W. and Skinner, J.L. 1972. Seasonal light trap collections of the two lined spittle bug in southern Georgia. *J. Econ. Entomol. 65(1)*: 110-14.

Beckham, C.M. 1970. Seasonal abundance of *Heliothis* spp. In Georgia piedmont. *J.Ga.Entomol. soc.* 5(3)138-142.

Bell, E. S., Jr. 1955. The relative attraction of certain commercially available electric lamps for hornworm moths. M.S. thesis at Va. Poly. Inst.

Belton, P. and Kempster, R.H. 1963. Some factors affecting the catches of Lepidoptera in light traps. *Canad. Entomol. 95*(8) 832-37.

Benkevich, V.I. 1959. Application of UV light traps in fighting noxious butterflies, Nauchn. *Dokl. Vysshel Shkoly,* No 3: 39-42.

Bensel, G. F. 1916. Control of the variegated cutworm in Ventura County. Calif. *Jour. Econ. Ent.* 9: 303-306.

Berger, R. S. 1966. Isolation, identification and synthesis of the sex attractant of the cabbage looper. *Ann. Ent. Soc. Amer.* 59: 767-771.

Berry, H.O., Blanc, F.L. and S.M. Klopfer. 1959. Pink bollworm detection in California. California Dept. Agric. Bull. 68: 211-18.

Betts, E. 1976. Forecasting infestations of tropical migrant pest: the desert locust and the African army worm. Pp. 113-114 in Raineg, R.C. (Ed.) *Insect Flight-287* pp. Oxford, Blackwell Scientific.

Birscoe, A.D. and Chitka, L. 2001. The evolution of colour vision in insects. Ann. Rev. Entomol. 46: 471-510.

Blair, B.W. and Catling, H.D. 1974. Out breaks of African army worm. *Spodoptera exempta* (Walker) (Lepidoptera: Noctuidae) in Rhodesia, South Africa, Botswana and South West Africa from February to April 1972. *Rhod. J. Agric. Res.* 12, 57-67.

Bogush, P. P. 1936. Some results of a study of light traps in Central Asia. *Bull. Ent. Res.* 27: 377-380.

Bogush, P. P. 1951. Application of light traps in study of insect population dynamics. *Entomol. Obozr. No.* 31: 609-28.

Bogush, P. P. 1958. Some results of collecting click beetles (Coleoptera: Elateridae) with light traps in middle Asia. *Rev. Ent. de URSS* 37 (2): 347-356.

Bogush, P. P. 1962. Flight dynamics of Ichneumonids to light traps in Bryansk in 1958. *Ibid.* 41: 572-5.

Bonnemaison, L. 1970. Comparative tests with coloured and light traps. *Anim.* 2(3): 391-422.

Borden, A. D. 1931. Some field observations on codling moth behavior. *Jour. Econ. Ent.* 24 (6): 1137-1145.

Bourne, A.I. 1936. Value of electric traps against orchard pests. *Mass. Agr. Expt. Sta. Bui.* 327, pp. 47-49.

Bowden, J. 1972. The significance of moon light in photoperiodic responses of insects. Bull. *Entomol. Res. 62:* 605-612.

Bowden, J. 1973. The influence of moon light on catches of insects in light trap in Africa. Part I. The moon and moon light. Bull. *Entomol. Res. 63* (1): 113-128.

Bowden, J. and Church, B.M. 1973. The influence of moon light on catches of insects in light traps in Africa. Part II. The effect of moon Phase on light trap catches. *Bull. Entomol. Res. 63*(1): 129-142.

Brader, L.M., Brader, l., Delalande, P. and P. Atger. 1968. Four years of observation of light traps in cotton growing in chad. *Cotton.Fibres Trop. Res. 23* (4): 63.

Briolini, G. and Celli, G. 1968. Results of the capture of Lepidoptera performed for a period of three years by a Pennsylvania- type light trap. *Boll. Ist. Entomol. Univ. studi. Bologna.* 29: 61-80.

Brown, E.S.; Betts, E. and R.C. Rainey. 1969 Seasonal changes in distribution of African army worm. *Spodoptera exempta* (Wik) (Lep.: Noctuidae) with special reference to Eastern Africa. *Bull. Ent. Res. 58.* 661-721.

Burkhardt, D. 1964. Colour discrimination in insect. In: Beament, J.E., Treherne and Wigglesworth, V.B. Ed. Advances in Insect Physiology vol. 2, Accd. Press. London.

Burks, B.D., Ross H.H. and T.H. Frison. 1938. An economical portable light for collecting nocturnal insects. *J. Econ. Entomol.* 31(2): 316-18.

Busck, A. 1917. The pink bollworm. *Jour. Agr. Res.* 9 (10): 354-370.

Butler, L. and Hunter, P. E. 1969. Reproduction and response to light in the black carpet beetle, *Attagenus megatoma* (Fabricius) (Coleóptera: dermestidae). *J. Georgia Ent. Soc.* 4 (4): 171-177.

Caffrey, D. J., and Worthley, L.H. 1927. Progress report in the investigations of the European corn borer. U.S. Dept. Agr. Bull. 1476,116 pp.

Calcote, V.R., Gentry, C.R. and G.W. Edwards. 1972. Comparison of two types of light trap in capturing moths of the pecan nut case bearer. *J. Econ. Entomol* 65(3): 933-34

Callahan, P. S. 1957. Oviposition response of the imago of the corn earworm, *Heliothis zea* (Boddie) to various wavelengths of light. Ann. Ent. Soc. Amer. 50 (5): 444-452.

Callahan, P.S., A.N. Sparks., J.W. Snow and W.W. Copeland. 1972. Corn ear worm moth: Vertical Nocturnal flight. *Environ. Entomol. 1*(4): 497-502.

Cantelo, W. W. and Smith, J.S. Jr. 1971. Collections of tobacco hornworm moths in traps equipped with one or four black light lamps baited with adult virgin females. *Jour. Econ. Ent.* 64 (2): 551-552.

Cantelo, W.W., Smith, J.S. Jr., Baumhover, A.H., Stanley, J.M. and T.J. Henneberry, **1972**. Suppression of an isolated population of the tobacco hornworm with black light traps, unbaited of baited with virgin female moths. *Environ. Entomol,* 1(2): 252-258

Cantelo, W.W., Smith. J.S. Jr., Beumbover, A.H., Stanley, J.M., Henneberry, T.J. and M.B. Peace. **1973.** Changes in the population levels of 17 insect's species during a 3½ years black light trapping Programme. *Environ. Entomol.* 2 (6): 1033-38.

Carruth, L.A. and Kerr, Jr. T.W. **1937.** Reduction of corn earworm moths and other insects to light traps. *J. Econ. Entomol.* 30: 297-305.

Chapin, J.B. and Callahan, P.S. **1967.** A list of the Noctuidae (Lepidoptera: Insecta) collection in the vicinity of Baton Rouge Louisiana. *Proc.La Acad Sci 30:* 39-48

Chavan, V.M. and Rao, R.S.N. **1981.** Insect pest of tobacco in north Bengal. *Indian Fmg.* July 81, pp13.

Chittenden, F. H. **1901.** Injurious moths attracted to lights in autumn. U.S. Dept. Agr., Div. Ent. N.S. Bui. 30: 85-86.

Chopra, R.L. **1928.** Annual report of the Entomologist to the Govt., Punjab Lyallpur for the year 1926-27. *Rep. Dept. Agric. Punjab 1* (2): 43-49.

Chu C.C., Jackson C.G., Alexander P.J., Karut K., and Henneberry T.J. **2003.** Plastic cup traps equipped with light-emitting diodes for monitoring adult *Bemisia tabaci* (Homopera: Aleyrodidae). *J Econ Entomol* **96**, 543–546.

Clemens, Brackinridge. **1859.** Instructions for collecting *Lepidoptera.* Smithsonian Inst. Bd. Regents Ann. Rpt. 173-200.

Click, P. A., and Graham, H. M. **1965.** Seasonal light-trap collections of lepidopterous cotton insects in South Texas. *Jour. Econ. Ent.* 58 (5): 880-882.

Collins, D. L. **1934.** Iris-pigment migration and its relation to behavior in the codling moth. *Jour. Expt. Zool.* 69: 165-197.

Collins, D. L. and Machado, W. **1937.** Effect of light traps on a codling moth infestation. *Jour. Econ. Ent.* 30 (3): 422-427.

Collins, D. L. and Nixon, M. W. **1930.** Responses to light of the bud moth and leafroller. N.Y. State Agr. Expt. Sta. (Geneva), Bui. 583, 32 pp.

Collins, D. L., and Machado, W. **1935.**Comments upon phototropism in the codling moth with reference to the physiology of the compound eyes. *Jour. Econ. Entom.,* 28: 103- 106

Collins, D.L. and Machado, W. **1943.** Reactions of the codling moth to artificial light and use of light traps in its control. *Ibid.* 36: 885-893.

Collins, D.L. and Machado, W.**1935.** Comments upon phototropism in the codling moth with reference to the physiology of the compound eye. *J. Econ. Entomol.* 28(1): 103-106.

Comstock, J. H. 1879. Fires, trap-lanterns, etc. In. USDA Div. Ent. Rpt. on cotton insects, pp. 262-275.

Cook, W. C. 1928. Light traps as indicators of cutworm moth population. *Canad. Ent.* 60 (5): 103-109.

Cook, W.C. 1930. Field studies of the Pale western cut worm (*Prorosagrotis orthogonia* morr.) *Bull. Montana Agric. Expt. Sta.*(225): 79.

Coon, B. F. 1968. Aphid trapping with black light lamps. *J. Econ. Ent.* 61 (1): 309-311.

Dalvania Dinesh kumar 2010. Role of light trap. Topic Seminar on M.Sc.Course (Ag. Ent. 510). C.P. College of Agriculture, S.D.A.U. Almora.

Dalziel, C. F.1951. Electric insect traps. Amer. Inst. Elect. Engin. Trans. 70: 1-5.

Davis, J. J.1935. The significance of supplementary controls in combating the codling moth. *Hoosier Hort.* 17 (3): 43-48.

Davis, J.J. 1953. Insects of Indiana for 1952. *Indiana Acad. Sci. Proc.* 62: 176-177.

Day, A. and Reid Jr. W. J. 1969. Response of adult southern potato wire worms to light traps. *J. Econ. Entomol.* 62(2): 314-18.

Deay, H. O. and Foster, G. H. 1944. Light traps reduce corn borer populations. In *Purdue Univ. Agr. Expt. Sta.Ann. Rpt.* 57: 48.

Deay, H. O. and Hartsock, J. G. 1960. The use of light traps to protect tobacco in southern Indiana from tobacco and tomato hornworms. *Ind. Acad. Sei. Proc.* 70: 137.

Deay, H. O. and Hartsock, J. G. 1961. The use of electric light traps as an insect control. In: Response of Insects to Induced Light. U.S. Dept. Agr., ARS 20-10, pp. 50-53.

Deay, H. O. Orem, M. T., and J. G. Taylor, 1953. Factors influencing flight of insects to ultra-violet light. Ann. Rpt. Purdue Agr. Expt. Sta. 66: 31-33.

Deay, H. O., Barrett, J. R., Jr., and J. G. Hartsock, 1965. Field studies of flight response of *Heliothis zea* to electric light traps, including radiation characteristics of lamps used.Proc.N.Centre. *Br.Entomol.Soc.Amer.20:* 109-16

Deay, H. O., Taylor, J. G., and E. A. Johnson, 1959. Preliminary results on the use of electric light traps to control insects in the home vegetable garden. Ent. Soc. Amer. No. Cent. Br. Proc. 14: 21-22.

Deay, H. O., Taylor, J. G., and J. R. Barrett, Jr. 1964. Light trap collections of corn earworm adults in Indiana in the years 1953-63. Ent. Soc. Amer. No. Cent. Br. Proc. 19: 45-53.

Deay, H.O. 1961. The use of electric trap as an insect control. P. 50-3 In response of insects to induced light. *USDA ARS.* 20-40.

Deay, H.O. and Taylor, J.G.1954. Preliminary report on the relative attractiveness of different heights of light traps to moths. *Proc. Indiana. Acad. Sci. 63:* 180-4.

DEB. Lylon, D.J. 1970. Catches of *Spodoptera littoralis.* (Boisd). In a light trap at Samaru. *Samaru. Res. Bull. 110:* 115-17

Debot, J.W., Jay D.L. and R.W. Ost. 1975. Light traps: Effect of modifications on catches of several species of Noctuidae and Arctiidae. *J.Econ. Entomol.* 68 (2): 186-88

Dethier, V.G. 1963. The physiology of insects senses, Methuen, London.

Dickerson, W.A., Gentry, C.R. and W.G. Mitchell. 1970. A rain free collecting container that separates desired Lepidoptera from smaller undesired insects in traps. *J. Econ. Entomol.* 63 (4): 1371.

Dina Nath. 1923. Preliminary observations on the attractions to light of moths of Sugarcane borers. *Rep. Proc. Vth Ent. Mtgs. Pusa*: 65-67.

Dirks, C.O. 1937. Biological studies of Maine moths by light trap methods. Maine Agr. Expt. Sta.Bull. 389,162 pp.

Ditman, L. P., and E. N. Cory, 1933. Corn earworm studies. Md. Agr. Expt. Sta. Bull. 348, 20 pp.

Drake,V.R. and Farrow, R.A. 1988. The influence of atmospheric structure and motions on insect migration. *Ann. Rev. Entomol.* 33: 183-210.

Dutt, H.L. 1919. The methods of control of *Agrotis ipsilon* in Bihar. Rep. *Proc. 3nl Entomol. Meet. Pusa II: 622-625, Calcutta,*

Edwin, W.K., D.P. Charles and K.R.John. 1965. An automatic sample changing device for light trap collecting, *J.Econ. Entomol. 58(1): 170.*

Ellertson, F.E. 1964. Trapping male Pleocoma with black light (Coleoptera: Scarabaeidae) *Pan. Pacific. Entomolgist. 40(3): 171-73.*

Ennerjeet B. 1988. Studies on the migratory behavior of *Agrotis ipsilon* (Hufnagal) based on the analysis of light trap catches. Ph.D. thesis Deptt. of Ento. JNKVV submitted to RDVV Jabalpur.

Essig, E. O.1930. A modern gnat trap. *Jour. Econ. Ent.* 23 (6): 997-999.

Eyer, J.R.1937. Ten years' experiments with codling moth bait traps, light traps, and trap bands. N. Mex. Agr. Expt. Sta. Bui. 253, 68 pp., illus.

Falcon, L. A., Van Den Bosch, Robert, Etzel, L. K., and others.1967. Light traps as detection devices for moths of cabbage looper and bollworm. *Calif. Agr.* 21 (7): 12-14.

Farrow R.A. and Daly J.C. 1987. Long range movements as an adaptive strategy in the genus Heliothis a review of its occurrence and detections in four pest species. *Aust. J. Zool.* 35: 1-24.

Ficht, G. A. and Anderson, S. A. 1942. Mercury lamps more efficient than mazda in corn borer moth traps. *In* Purdue Univ. Agr. Expt. Sta. Ann. Rpt. 55: 53.

Ficht, G. A., Hienton, T. E. and J. M. Fore, 1940. Use of electric light traps in the control of the European corn borer. *Agr. Engin.* 21 (3): 87-89.

Ficht, G.A. and Hienton, T. E. 1939. Studies on the flight of European corn borer moths to light traps: a progress report. *J. Econ. Entomol. 32* (4)- 520-526.

Ficht, G.A. and Hienton, T. E. 1941. Some more important factors governing the flight of European corn borer moths to electric traps. *Ibid: 34(5):* 599-604.

Fisk, F.W. and Perezperez, R.1969. Flight activity periods of the sugarcane borer *Diatraea saccharalis,* in Puerto Rico. *J.Agric. Univ. Puerto Rico. 53(2):* 93-9.

Fletcher, T.B. 1916. Report of Imperial Entomologists. Rep. Agr. Res.Inst. and College, Pusa, 1915-16: 58-77 Calcutta.

Flint, G.J. 1969. Light trap catches of Lepidoptera in 1967 near the Abbey of Sion. *Entomol. Ber. 29 (7):* 211-22.

Fore, P. L. 1969. Response of fish to light. Ph.D. dissertation, Southern Illinois University, 85 p.

Frost, S. W. 1949. The Dimond back moth in Pennsylvania. *J.Econ. Entomol.42(4):* 681-682.

Frost, S. W. 1952. Light trap for insect collection survey and control. *Pan Agric. Exp. Stat. Bull.* 550,32pp.

Frost, S. W. 1952. Miridae from light traps. J. N.Y. Ent. Soc. 60: 237-240.

Frost, S. W. 1953. Response of insects to black and white light *J.Econ. Entomol. 46(2):* 376-7.

Frost, S. W. 1954. Response of insects to black and white light. *Journal of Economic Entomology,* 47(2): 275-278.

Frost, S. W. 1955. Response of insects to ultraviolet light. *Ibid.* 48(2): 155-156.

Frost, S.W. 1957a. The Pennsylvania insect light trap. *Jour. Econ. Ent.* 50 (3): 287-292.

Frost, S.W. 1958. Insects attracted to light traps placed at different height. *Jour. Econ. Ent.* 51(4): 550-551.

Frost, S.W. 1964a. Insects taken in light traps at the Archbold Biological Station, Highlands County, Fla. *Fla. Ent.* 47 (2): 129-161.

Frost, S. W. 1964b. Killing agents and containers for use with insect light traps. *Ent. News* 75 (6): 163-166.

Frost, S.W. 1966. Addition to Florida insects taken in light traps. *Fla. Ent.* 49 (4): 243-251.

Fu, S.F., C.S. Wan and C.Y. Tsao. 1959. A black light fluorescent lamp used in predicting the appearance of several important cotton insects. *East China J. Agric. Sci. 1:* 44-5.

Gambrell, F. L., Mendall, S. C, and E. H. Smith, 1942. A destructive European insect new to the United States. *Jour. Econ. Ent.* 35 (2): 289.

Ganguli, J; Ganguli, R.N. and S.M. Vaishampayan, 1997. Identity of migratory phase of cabbage semilooper *Plusia orichalcea* (Fabr) on the basis of neurosecretory cells of brain. *J. Insect Sci.* 10 (2): 172-173.

Ganguli, J; Ganguli, R.N. Vaishampayan, S.M. and R. Verma, 1989. Identification of migratory phase of cabbage semilooper *Plusia orichalcea* Fabr. (Lepidopetera:

Noctidae) on the basis of reproductive development and fat body content. *J. Adv. Zool.* 10(2): 89-94.

Gaskin, D.E. 1970. Analysis of light Catches of Lepidoptera from palmerston North, New Zealand in 1966-68. *NZJ. Sci. 13* (3): 482-99.

Gehring, R. D. and Madsen, H. F. 1963. Some aspects of the mating and oviposition behavior of the codling moth. *Jour. Econ. Ent.* 56 (2): 140-143.

Geier, P. W.1960. Physiological age of codling moth females caught in bait and light traps. *Nature* 4714: 709.

General Electric Co. 1964. Fluorescent lamps performance data. 602-11001, Tab #3. Loose leaf np.

Gentry, C.R., Dickerson, W. A. Jr. and J.M. Stanley. 1971. Population and mating of adult tobacco budworms and corn ear worm in north-west Florida indicated by traps. *Idid. 64*(1): 335-8.

Gentry, C.R., F.R. Lawson, C.M. Knott. J.M. Stanley and J.J.Lam, Jr. 1967. Control of hornworms by trapping with black light and stalk cutting in North Carolina. *J. Econ. Entomol. 60*: 1437-42.

Georgia Agricultural Experiment Station. 1938. Miscellaneous insects at light traps. Ann. Rpt. 1937-38: 69-70.

Gillette, C. P. 1897. A successful lantern trap. Ninth Ann. Mtg. Assoc. Econ. Ent. Proc, U.S. Dept.Agr. Div. Ent. Bui. 9: 75-76.

Girardeau, M. F., Stanley, J. M., and La Hue, D. W. 1952. Preliminary report on light traps for catching night-flying insects. Ga. Coastal Plain Expt. Sta. Tech. Mimeo. Paper No. 5, 27 pp.

Glick, P.A. and Graham, H.M. 1961. Early season collections of three cotton insects by argon glow lamp and black light traps. *J.Econ. Entomol 54*(6): 1254.

Glick, P.A. and Graham, H.M. 1965. Seasonal light trap collection of lepidopterous cotton insects in South Texas. *Ibid. 58*(2): 880-882.

Glick, P.A. and Hollingsworth, J.P. 1954. Response of the Pink bollworm moths to certain ultraviolet and visible radiation. *J.Econ. Entomol 47*(1): 81-6.

Glick, P. A. and Hollingsworth, J.P. 1955. Response of the pink bollworm and other cotton insects to certain ultraviolet and visible radiation. *J.Econ. Entomol 48*: 173-177.

Glick, P.A., Hollingsworth, J.P. and W.J. Eitel. 1956. Further studies on the attraction of pink bollworm moths to ultraviolet and visible radiation. Ibid. 49(2): 158-161.

Glover, Townend.1865. U.S. Commr. agr. rpt. of entomolgist. pp. 540-564.

Goldsmith, T.H. 1961. The colour vision of insects. In Light and Life. (W. D. Mc Ekoy and B. Glass, eds.), pp.771-794. John Hopkins Press, Baltimore, Md.

Graham, H. M. and Bilhngsley, C. H. 1964. Evaluation of a blower attachment for light traps. *Jour. Econ. Ent.* 57 (1)-169-170.

Graham, H. M., Hollingsworth, J. P., Lukefahr, M. J., and J. R. Lianes, 1971. Effects of a high density of black light traps on populations of the corn earworm in corn. U.S. Dept. Agr. Prod. Res. Rpt. 127, 24 pp.

Graham, H.M., Glick, P.A. and D.E. Martin. 1964. Nocturnal activity of adults of six Lepidopterous pests of cotton as indicated by light trap collections. *Ann. Entomol. Soc. Amer.* 57(2): 328-32.

Graham, H.M., Glick, P.A. and J.P. Hollingsworth. 1961. Effective range of argon glow lamp survey traps for pink boll worm. *J.Econ Entomol.* 54(4): 788-9.

Graham, H.M., Glick, P.A., Ouye, M.T. and D.F. Martin. 1965. Mating frequency of female pink bollworms collected from light traps. *Ibid. 58(5):* 595-96.

Gross, A. O.1913. The reactions of arthropods to monochromatic lights of equal intensities. *Jour.Expt. Zool.* 14 (4): 467-514.

Groves, J.R. 1955. A comparison of bait and light traps for catching codling moths, *Cydia pomonella* (L). *Ann. Rept. East Malling Res. Sta. (England):* 146-8.

Gryse, J. J. de.1933. Note on a new light trap. Ent. Soc. Ont. Ann. Rpt. 64: 55-57.

Gui, H.L.; Porter, L.C. and C.F. Paraideaux 1942. Response of insects to colour, intensity and distribution of light. Am. Soc. Agril. Engineers. J. 23: 51-58.

Haggis, M.J. 1972. Light trap catches of *Spodoptera exempta* (Walk.) in relation to wind direction. *East. Afr. Agric. For. J.* 37(2): 100-108.

Hallock, H.C.1932. Traps for the Asiatic garden beetle. Jour. Econ. Ent. 25: 407-411.

Hallock, H.C.1936. Recent developments in the use of electric light traps to catch the Asiatic garden beetle. *Jour. N.Y. Ent. Soc.* 44 (4): 261-279.

Hamilton, D.W. and Steiner, L.F. 1939. Light trap and codling moth. *J.Econ Entomol* 32(6): 867-72.

Hanna, H.M. 1970. Studies on catches of coleopteran in a light trap at Assiut. *Bull. Soc. Entomol. Egypte.* 53: 591-613.

Hanna, H.M. and Ibrahim, E. 1968. On the time flight of certain nocturnal Lepidoptera as measured by a light trap. *Ibid.* 52: 535-45.

Harcourt, D. G. 1957. Biology of the diamond back moth, *Plutella maculipennis* (Curtis) (Lepidoptera: Plutellidae) in Eastern Ontario. II Life-history, behavior, and host relationships. *Canad. Ent.* 89: 554-564.

Harcourt, D. G., and Cass, L. M. 1958. A controlled-interval light trap *for micro lepidoptera, Canad. Ent.* 90: 617-622.

Harding, C.H. Jr., Hartsock, Jr. J.G.. and G.G. Rohwer. 1966. Black light traps standards for general insect survery. *Bull. Entomol. Soc. Amer.* 12: 31-32.

Hardwick, D.F. 1965. The corn earworm complex. *Memoirs Entomol. Soc. Canada* 40: 1-247.

Hardwick, D.F. 1970. A generic revision of the North American *Heliothidinae* (Lepidoptera: Noctuidae) *Memoirs Entomol. Soc. Canada* 73: 1-59.

Hardwick, D.F. 1968. A brief review of the principles of light trap design with a description of an efficient trap for collecting noctuid moth. *J. Lepidopterists Soc.* 22(2): 65-75.

Harinkhere, J.P.; Kandalkar, V.S. and A.K. Bhowmick, 1998. Seasonal abundance and association of light trap catches with field incidence of rice leaffolder (*Cnaphalocrocis medinalis* Guenee). *Oryza.* 35: 91-92.

Harling, J. 1968. Meteorological factors affecting the activity of night flying macro Lepidoptera. *Entomologist* 101(1259): 83-93.

Harrell, E. A., Young, J. R., and A. C. Cox, 1967. Fan vs. gravity light traps for collecting several species of *lepidoptera. Jour. Econ. Ent.* 58(5): 1010-1011.

Harrendorf, K. 1959. Occurance and relative abundance of certain noctuid moths in north-west Arkansas Fall. 1957. *J. Kanas. Entomol. Soc.* 32(1): 41-44.

Harris, T. M. 1921. On a mode of destroying moths. Mass. Agr. Rep.Jour. 6: 392-93.

Harry, Catz, 2010. Information from internet.

Hartstack, A. W., Jr., and Hollingsworth, J. P.1968. An automatic device for dividing and packaging light trap insect catches according to time intervals. U.S. Dept. Agr., Agr. Res. Serv. ARS 42-146, 14pp.

Hartstack, A.W. Jr., Henson J.L., Witzj, J.A., Jackman, J.A. Hollingsworth, J.P. and R.E. Frisbie 1977. The Texas program for forecasting *Heliothis* spp. *Infestations on cotton Proceedings 1977. Beltwide cotton Production Research Conference, Atlanta, G.A.*

Hartstack, A.W. Jr., Hollingsworth J.P. and D.A. Lindquist. 1968. A technique for measuring trapping efficiency of electric insect trap. *J. Eco Entomol.* 61(2): 546-52.

Hartstack, A.W. Jr., Hollingsworth, J.P., Ridgway, R.L. and R.R. Coppledge. 1973. A population dynamics study of the bollworm and the tobacco budworm with light traps *Environ. Entomol.* 2 (2): 244-5.

Hartstack, A.W., Jr., Hollingsworth, J.P. Ridgway, R.L. and H.H. Hustt. 1971. Determination of trap spacing required to control an insect population. *J. Eco Entomol.* 65 (5): 1090-1100.

Hartstack, J. G. 1961. Relation of light intensity to insect response. *In* Response of Insects to Induced Light, U.S. Dept. Agr., Agr. Res. Serv. ARS 20-10, pp. 26-32.

Hartstack, J.G., Deay, H.D. and J.R. Barrett., Jr. 1966. Practical application of insects attraction in the use of light traps. *Bull. Entomol. Soc. Amer.* 12(4): 375-77.

Hassanein, M.A. 1972. Abundance and population density of three lepidopterous insects in upper Egypt. (Lepidoptera: Noctuidae) *Bull. Soc. Entomol.* Egypt 55: 79-83.

Hawley, I. M.1936. The construction and use of light traps designed to catch the Asiatic garden beetle. *U.S. Dept. Agr. Bur. Ent. Plant Quar.* E-385, 6 pp.

Hays, S.B. 1968. Adult hornworm populations and degree of infestation on tobacco in relation to community wise grower use of black light traps. *J. Econ. Entomol. 61(3):* 613-17.

Heidinger Roy C. (1971). Use of ultraviolet light to increase the availability of aerial insects to caged Bluegill Sunfish. Publication No. 41 Fisheries and Illinois Aquaculture Center. *The Progressive Fish-Culturist*, Vol. 33, Issue 4 (October 1971). pp. 187-192.

Hendricks, D. E.1968. Use of virgin female tobacco budworms to increase catch of males in black light traps and evidence that trap location and wind influence catch. *Jour. Econ. Ent.* 61 (6): 1581-1585.

Hendricks, D.E.; Graham, H.M. and J.R. Raulston, 1973. Dispersal of sterile tobacco budworm from release points in North eastern Mexico and Southern Texas *Environ. Entomol.*: 1085-1088.

Henneberry, T.J. and Howland, A.F. 1966. Response of male cabbage loppers to black light with or without the presence of the female sex pheromone. *J.Econ. Entomol. 59(3)*: 623-6.

Henneberry, T.J., Howland, A.F. and W.W. Wolf 1967. Combinations of black light and virgin females as attractants of Cabbage looper moths. *Ibid. 60*: 152-6.

Herms, W. B. 1929. A field test of the effect of artificial light on the behavior of the codling moth, *Carpocapsa pomonella* (Linn.). *Jour. Econ. Ent.* 22 (1): 78-88.

Herms, W. B. 1932a. A light trap with suction fan. *J. Am. Soc. Agric. Engin.* 13: 292

Herms, W. B. 1932b. Deterrent effect of artificial light on the codling moth. Univ. of Calif. Hilgardia 7 (9): 263-280.

Herms, W. B. 1947. Some problems in the use of artificial light in crop production. Calif. Agr.Expt. Sta. Hilgardia 17 (10): 359-375.

Herms, W. B. and Burgess, R. W. 1928. Combined light and suction fan trap for insects. Elec. West 60 (4): 204-205.

Herms, W. B. and Ellsworth, J. K. 1934. Field tests of the efficacy of coloured light in trapping insect pests. *Jour. Econ. Ent.* 27 (5): 1055-1067.

Herms, W. B., and Ellsworth, J. K. 1935. The use of coloured light in electrocuting traps for the control of the grape leafhopper. *Agr. Engin.* 16(5): 183-186.

Hervey, G. E. R., and Palm, C. E. 1935. A preliminary report on the response of the European corn borer to light. *Jour. Econ. Ent.* 28 (4): 670-675.

Hienton T.E. 1974. Summary of investigations of electric insect traps. USDA technical bulletin No.1498.

Hoffman, D.J., Lawson, F.R. and B. Peace. 1966 Attraction of black light traps with virgin female tobacco hornworm moths. *Ibid. 59(1):* 809-11.

Hofmaster, R. N. 1966. Light trap collections of insects and its value to agriculture. Va. Veg. Growers News, Mar. 1962, p. 2.

Hollingsworth, J. P. and Briggs, C. P. 1960. A transistorized power supply and automatic control unit for battery operation of survey-type electric insect traps. U.S. Dept. Agr., Agr. Res. Serv. ARS 42-38,16 pp.

Hollingsworth, J. P. Wright, R. L., and D. A. Lindquist, 1964. Radiant-energy attractants for insects. Agr. Engin. 45 (6): 314-317.

Hollingsworth, J.P. 1961. Relation of wave- length to insect response. *USDA, ARS(Ser)*20-10: 9-25.

Hollingsworth, J.P., Briggs, C.P., Glick, P.A. and H.M. Graham. 1961. Some factors influencing light trap collection *J.Econ. Entomol. 54*(2): 305-8.

Hollingsworth, J.P., Hartstack, A.W. Jr. and D.A. Lindquist. 1968. Influence of near ultraviolet output of attractant lamps on catches of insects by light trap *Ibid*. 61(2): 515-521.

Hollingsworth, J.P., Hartstack, J.G. and J.M. Stanley. 1963. Electric insect trap for survey purposes. USDA. ARS (Ser). 1(2): 42-3.

Hollingsworth, J.P., Wright, R.L., and D.A. Lindquist, 1964. Spectral response characteristics of the boll weevil. *J. Econ. Ent.* 57 (1): 38-41.

Holzman, R. W. 1961. Collecting Sphingidae with a mercury vapor lamp. *J. Lepid. Soc.* 15 (3): 191-194.

Horsfall, W. R. 1962. Trap for separating collections of insects by interval. *Jour. Econ. Ent.* 55 (5): 808-811.

Horsfall, W. R. and Tuller, A. V. 1942. An apparatus for obtaining interval collection of insects. *Ent. News* 53: 253-258.

Hosny, M.M.; Iss-Hak, R. R.; Nasir, El Sayed; El Deeb, Y.A.; Critchiey, B.R. Topper,C. and D.G. Campion 1979. Mass Trapping for the control of the Egiptian cotton leafworm *Spodoptera littoralis* (Boisd) in Egypt. In: Proceed. 1979 British Crop Protection Conference, Pest and disease. 395-400 pp.

Howard, L. O. 1897. Insects affecting the cotton plant. *U.S. Dept. Agr. F. Bull.* 47, 32 pp.

Howard, L.O. 1900. The principal insects affecting the tobacco plant. *U.S. Dept. Agr. F. Bull.* 120, 32 pp.

Howland, A. F., and Wolf, W. W. 1967a. Combinations of black light and virgin females as attractants to cabbage looper moths. *Jour. Econ. Ent.* 60 (1): 152-156.

Hurst, G.W. 1965a. Nocturnal activity of insects as indicated by light traps. *Bull. Soc. Entomol. Egypt* 40: 463-479.

Hurst, G.W. 1965b. *Laphygma exigua* immigrations into the British Isles 1947-63. *Int. J.Biochem Biomet*, 2: 21-28.

Hussain, M.A. 1930. Entomology *Rep. Dept. Agril. Punjab. 1928-29*, (1): 53-63.

Hussain, M.A., Khan, M.H. and G. Ram. 1934. Studies of *Platyedra gossypielle* Saunders, the pink boll worm of cotton in the Punjab. Part III. Phototropic responsa of P. gossypielle. *Indi.J.Agric. Sci. 1:* 261-9.

Hutchins, R.E. 1940. Insect activity at a light trap during varios periods of the night. *J.Econ.Entomol. 33(4):* 654

Ishikura, H. 1967. Assessment of the field population of rice stem borer moth by light trap. In: *Procedings of a symposium on the major insect pests of rice plant 14-18 Sep. 1964,* Los. Banos. Laguna Phillippines, Jogns. Hopkins press: Baltimore, Md.169-179.

Jalas, I. 1960. A lightly built easily transportable light trap for Catching Lepidoptera. *Ann. Entomol. Fennici. 26(1):* 44-50.

Javeri, T.N. 1921. Notes on 'Katra'(Hairy caterpillars) and their controlling measures. Rep. *Prod. 4ᵗʰ Ent. Mtgs. Pusa:* 98-100.

Jeon J.H., Oh M.S., Cho K.S., and Lee H.S. 2012. Phototacic response of the rice weevil, *Sitophilus oryzae* Linnaeus (Coleoptera: Curculionidae), to light emitting diodes. *J. Korean Soc Appl Biol Chem* **55**, 35–39.

Johnson, C.G. 1950. A suction trap for small airborne insects which automatically segregates the catch into successive hourly samples. *Ann. Appl. Biol.* 37 (1): 80-91.

Johnson, C.G. 1963b. The origin of flight in insects. *Proc. Royal Ent. Soc. London* (c) 28: 26-27.

Johnson, C.G. 1969. Migration and dispersal of insects by flights. *Methuen and Co. Ltd., London,* pp- 763.

Jones, G.A. and Richard, T.1970. Effect of an area programme using black light trap to control populations of tobacco horn worms and tomato horn worms in Kentucky. *J. Econ. Entomol. 63(4):* 1187-94.

Jong, D. J. de, and Pol, P. H. Van de. 1955. Use of light traps to determine the flight of the fruit moth and the leaf roller. De Fruitteelt 45 (8): 200-202,

Juliet, J.A. 1963. A comparison of four types of traps used for capturing flying insects. *Canadian, J.Zool. 41(2):* 219-23

Kelsheimer, E. G. 1935. Responses of European corn borer moths to coloured lights. *Ohio Jour. Sei.* 35 (1): 17-28.

Kent, T. E., III. 1958. A laboratory study of the response of the tobacco hornworm moth to ultraviolet radiation. *M.S. thesis at Va. Poly. Inst.*

Kerketta, M.S.; Dubey, A.K. and U.K. Kaushik, 1990. Light trap studies of two rice planthoppers, *Nilaparvata lugens* and *Sogatella furcifera* in relation to field population. *Oryza.* 27: 503-506

Khan, R.H. 1983. Studies on the common predatory and parasitic species of insects collected on light trap at Jabalpur. *M.Sc. (Ag.) thesis,* JNKVV Jabalpur, M.P.

Khattab, A.A.S. 1975. Moths population of *Spodoptera littoralis* as attracted by an ultra-violet light trap. *Agril. Res. Review,* 53 (1): 13-19

Killough, R. A. 1961. The relative attractiveness of electromagnetic energy to nocturnal insects. *Ph.D. thesis.* Purdue Univ., Lafayette, Ind.

Kim, M.G., Yang, J.Y., Chung, N.H. and H.S.Lee. 2012. Photo responce of tobacco whitefly, *Bemicia tabaci* G. (Hemiptera: Aleyrodidae), to Light- emmiting Diodes. *J. Koreon Soc. Appl. Biol. Chem.* 55: 567-569.

King, E. W., and Hind, A. T. 1960. Activity and abundance in insect light trap sampling. *Ann. Ent. Soc. Amer.* 53 (4): 524-529.

King, E.W., D.P. Charles and J.K. Reed. 1965. An automatic sample changing device for light trap collecting. *J.Econ. Entomol. 58(1):* 170-2.

Kishaba, A. N., Wolf, W. W., Toba, H. H., Howland, A.F. and T.Gibson 1970. Light and synthetic pheromone as attractants for male cabbage loopers. *J.Econ. Ent.* 63 (5): 1417-1420.

Klocker, Alb. 1903. An apparatus for the capture of *lepidoptera* at light. *Ent. Medd., Ser.* 2 (1): 52-53.

Knaggs, H.G. 1866. The new American moth trap. *Ent. Monthly Mag.* 2: 199-202.

Knowlton, G. F. 1964. Nine years of corn earworm moth trap light collections in Northern Utah, 1956-1964. Utah State Univ. Agr. Ext. *Ent. Mimeo. Ser.* 31: 7

Koch, L.E. 1953. Light trap catches of two species of rice stem borer moth near Wyndham. Western, Australian, *Natur.9(1):* 12-14.

Koloostian, G. H., and Wolf, W. W. 1968. Attraction of pear psylla to black light. *J. Econ. Ent. 61* (1): 145-147.

Kovitvadhi, K., and Cántelo, W. W. 1966. The use of light traps to measure relative population densities of the rice gall midge, *Pachydiplosis oryzae,* and other insects of rice. *Internatl. Rice Comm. News Let.* 15(2): 16-19.

Kugerberg, Otto. 1968. Mating and ovary studies on *Cerapterix graminis* L. (Lepidoptera: Noctuidae). Collected from a light trap. *Opuscula. Entomol.* 33(1/2): 107-10.

Kunerth, W. 1919. Lighting for country homes and village communities. Iowa State Coll. *Engin.Expt. Sta. Bull.* 55, 32 pp.

Lam, J. J., Stanley, J. M., Knott, C. M., and A. H. Baumhover, 1968. Suppression of nocturnal tobacco insect populations with black light traps. *Am.Soc. Agr. Engin. Trans.* 11 (5): 611-612.

Lam, J. K., Jr. 1964. Response of hornworm moths to monochromatic radiation in the visible and near ultraviolet spectrum. M.S. thesis at Va. Poly. Inst.

Lam, J.K. Jr. and Stewart, P.A. 1969. Modified traps using black light lamps to capture nocturnal tobacco insects. *J. Econ. Entomol 62(6):* 1378-81.

Lansburg, I. 1960 Notes on corixidae occurring at black light traps (Hemiptera: Heteroptera) *Entomol.News.71(9):* 244.

Lawson, F. R., Gentry, C. R., and J. M. Stanley, 1963. Effect of light traps on hornworm populations in large areas. U.S. Dept. Agr.,*Agr. Res. Serv. ARS* 33-91, 18 pp.

Lawson, F. R., Gentry, C. R., and J. M.Stanley, 1966. Experiments on the control of insect populations with light traps. U.S. Dept.Agr., *Agr. Res. Serv. ARS* 33-110, pp. 194-214.

Lawson, F. R., Knott, C. M., and others. 1967. Control of hornworms by trapping with black light and stalk cutting in North Carolina. *J. Econ. Entomol. 60* (5): 1437-1442.

Lingren, P.D. and Wolf, W.W. 1982. Nocturnal activity of the tobacco budworm and other insect. In: The role of biometeorology in Integrated Pest Management (Eds. J.L. Hartfield and I.*J. Thompson Acad. Press NY* pp: 205-222.

Lipa, J.J. 1972. Visual insects attractants and their use in plant protection. *Postepy. Nauk. Roln. 19*(9): 24-44

Liu, L.C. 1993. Application of double-wave black light trap in pest insect forecasting. *Ento. Knowledge*, 30(3): 161-166.

Loftin, V. E., McKinney, K. P., and W. H. Hanson, 1921. Report on investigations of the pink bollworm of the cotton in Mexico. U.S.Dept. Agr. Bull. 918, 64 pp.

Lubbock, J.1882. On the senses of ants. In: Ants, bees, and wasps. The International Scientific Series D., Appleton and Co. London, pp. 182-207.

Luckiesh, M. 1946. Applications of germicidal, erythemal and infrared energy, New York, D. Van Nostrand Co., 463 pp.

Madsen, H. F. 1967. Codling moth attractants. *Pest Art. and News Sum.*13 (4): 333-344.

Madsen, H.F. 1962. Black light traps help in determine flights of codling moth and other deciduous fruit pests. *California Agric. 16*(2): 12-13.

Mahobe, J. and Vaishampayan, S.M. 1990a. Studies on the seasonal variation in the reproductive potential and ovary development of soybean semilooper *Plusia chalcytes* Fabr. *Geobios* (17): 237-241.

Mahobe, J. and Vaishampayan, S.M. 1990b. Seasonal variation in the reproductive potential of female of *Plusia orichalcea* Collected from light trap. *Proceed. Nat. Acad. Sci. India. 60* (B): IV: 373-76.

Mallicky, H. 1967. Causes and extent of migration of butterflies. Umachan.Wiss. *Tech. 67*: 501.

Mangat, B. S., and Apple, J. W. 1964. Heat units and light trap collections as methods for timing corn earworm insecticidal treatments. *Ent. Soc. Amer. No. Cent. Br. Proc.* 19: 108-109.

Marshall, G. E., and Hienton, T. E. 1935. Light traps for codling moth control. *Agr. Engin.* 16: 365-368.

Marshall, G. E., and Hienton, T. E. 1938. The kind of radiation most attractive to the codling moth. *Jour. Econ. Ent.* 31(3): 360-366.

Marten (1956). *Entomol. Z.* 66: 121-133.

Martin, C. H., and Houser, J. S. 1941. Numbers of *Heliothis armigera* (Hub.) and two other moths captured at light traps. *Jour. Econ. Ent.* 34 (4): 555-559.

Maxwell-Lefroy, H. 1906. The pink boll-worm. In: Indian Insect Pests. Off. Supt. Govt. Printing, India, pp. 93-96.

Mazokhin Porshnyakov G.A. 1969. *Insect Vision.* Platinum Press, New York.

Mazokhin Porshnyakov G.A. 1958. Construction and use of UV light traps. *Entomol. Obozr.* 37: 464-71.

Mazokhin Porshnyakov, G. A. 1956. Application of the ultraviolet rays for control of the June bug. Zool. J. 35 (9): 1356-1361.

Mazokhin Porshnyakov, G.A. 1954. Use of UV, light in the fight against harmful insects.. In: 'Tr.Soveschch. Porybov," Moscow, p. 404-6.

Mazokhin Porshnyakov, G.A. 1956. Comparison of insect attracting effectiveness of radiations having different spectral compositions. *Enomol. Obozr* 35: 752-759.

Mazokhin Porshnyakov, G.A. 1960. Why insects fly towards the light. *Enomol. Obozr* 39: 52-58.

Mazokhin Porshnyakov, G.A. 1964. Methods for the study of insect colour vision. *Enomol. Obozr.* 43: 503-523.

McFadden, M.W. and J.J. Lam., Jr. 1968. Influence of population level and traps spacing on capture of tobacco hornworm moth in black light traps with virgin females. *J.Econ. Entomol.* 61(5): 1150-52.

McLelland, C.K., and C.A. Sahr, 1911. Cultural methods for controlling the cotton bollworm. *Hawaii Agr. Expt. Sta.Press Bull.* 32.

McNeill, Jerome. 1889. An insect trap to be used with the electric light. *Amer. Nat.* 23: 268-270.

Medler, J.T. and P.W. Smith. 1960. Membracidae attracted to black light. *Ibid.* 53(1): 173-74.

Mehrhof, F.E., and E.R.Van Leeuwen, 1930. An electrical trap for killing Japanese beetles. *Jour. Econ. Ent.* 23 (1): 275-278.

Menear, J.R. 1961. Response of tobacco and tomato hornworm moths to monochromatic radiation in the near ultra-violet. *M.S. thesis* at Va. Poly. Inst.

Merkl, M.E., and T.R. Pfrimmer, 1955. Light-trap investigations at Stoneville, Miss., and Tallulah, La., during 1954. *Jour. Econ. Ent.* 48 (6): 740-741.

Mesch, H. 1966. Light trap serving plant protection. *Entomologische. Berichte.* 3: 9-19.

Mesch, H.1965. Experiences with light traps for the warning service. *Beitr. Ent.* 15 (1-2): 139-154.

Meshram, P.B.; Pathak, S.C. and Jamaluddin 1990. Population dynamics and seasonal abundance of some forest insect pests (nursery stage) through light trap. *Indian-Forester.* 16: 494-503

Meurer, J.J. 1957. Survey of Heteroptera captured with a light trap in Heemstede (near Haarlem) in 1955. *Ent. Ber.* 17 (5): 80-96. [In Dutch, English summary, p. 96.]

Michelbarger, A.E., and Essig, E.O. 1938. Caterpillars attacking tomatoes. Calif. *Agr. Expt. Sta. Bui.* 625, 42 pp., illus.

Mikkola, K. and P.Salmensuu, 1965. Migration of *Laphygma exigua* Hb. (Lepidoptera: Noctuidae) in Northern Europe in 1964. *Ann. Zoo. Fenn.* 2: 124-39.

Milne, D. 1927. Entomology. Rep. Dept. Agric. Punjab 1925-26. (1): 49-54.

Milne, L.J., and Milne, M.J.1944. Selection of coloured lights by night-flying insects. *Ent. Amer,* (n.s.) 24 (2): 21-86.

Mirzayans, H. and Hodjat, S.H. 1971. List and relative abundance of moths caught by light traps. In 1970 from alfalfa field in Evoin. (Tehxan). *Pl. Pests Dis. Res. Pro.* 2(5): 45-51.Tehran.

Mohan, S; and Janarthanan, R. 1985. Effect of light trap on the incidence of yellow rice borer (*Scirpophaga incertulas* Wlk.) in trap zone and field. *Oryza.* 22: 61-64.

Mohanraj, D.R. Janarthanan and S. Suresh (1989). Response pf rice pests to mercury vapour light and black light traps. *Intl. Rice Res. Newsletr.* 14 (4): 37

Mohanraj, D; Janarthanan, R; and S. Suresh, 1989. Sex and reproductive status of rice stem borers and leaffolders attracted to black light trap. *International Rice Research Newsletter.* 14: 37.

Morgan, A. C, and S. C.Lyon, 1928. Notes on amyl salicylate as an attractant to the tobacco hornworm moth. *Jour. Econ. Ent.* 21 (1): 189-191.

Morgan, H. A. 1897. Lights in collecting bollworm moths. *In* Rpt. Ent. La. Agr. Expt. Sta. Bull. Second Ser. 48, p. 155.

Nagel, R. H., and Granovsky, A. A.1947. A turntable light trap for taking insects over regulated periods. *Jour. Econ. Ent.* 40(4): 583-586.

Nakamoto Y., Kuba H. 2004. The effectiveness of a green light emitting diode (LED) trap at capturing the West Indian sweet potato weevil, *Euscepes postfasciatus* (Fairmaire) (Coleoptera: Curculionidae) in a sweet potato field. *Appl Entomol Zool* 39(3): 491–495

Nakazime, S.Y., Nakashima and E.Yamaimato. 1968. Effect of group light trapping on control of fruit piercing moths in citrus groves (in Japan). In: (Japanese). *Proc. Ass.Pl. Prot. Kyushu.* 14: 70-71.

Nandihalli, B.P.; Patil, B.V. and P. Huger, 1990. Influence of weather factors on the light-trap catches of yellow rice borer under Tungabhadra project area. *Indian Journal of Ecology.* 17: 90-93.

Nandihalli, B.P.; Patil, B.V. and P. Huger, 1991. Influence of weather factors on the light trap catches of the sorghum stem borer. *J. Maha. Agri.Uni.*16: 179-181.

Nasar, E.S.A. and Hosney, H.M. 1980. The possible north bound movement of *Agrotis ipsilon* moths in the spring as indicated by catches in a series of light traps. Pl. Prot. Instt. Mini of Agril. Dokki. Egypt, pp. 36-39.

Nath Vishwa, (1983). White grub and its management. In: Principles and concepts of Integrated Pest Management. Edt.: R. K. Agrawal *et al.,* ICAR publication pp.135

Nath Vishwa; Shrivastava, A.K.; Kawal Dhari; Rajendra Singh and R.S. Verma 1978. Effect of light trap on the the beetles of white grub *Holotrichia consanguinea* Balncherd. *Indian J. Ent.* 40 (4): 465

Nelson, S. O. 1967. Electromagnetic energy in "pest control." (W. W. Kilgore and R. L. Doutt,eds.), 477 pp. Academic Press, New York, N.Y.

Nemec, S.J. 1969. Use of artificial lighting to reduce *Heliothis* spp. Populations in cotton fields. *J.Econ.Entomol.* 62: 1138-1140.

Nemec, S.J. 1971. Effect of lunar phase on light trap collections and populations of boll worm moths. *Ibid.* 64(4): 860-64.

Newcomb, D. D. 1967. Comparative behavior of adult *Heliothis zea* and *Heliothis virescens* to light. M.S. thesis at Tex. A and M Univ.

Newcomer, E.J., Yothers, M.A., and W.D. Whitcomb, 1931. Control of the codling moth in the Pacific Northwest. U.S. Dept. Agr. F. Bull.1326, 26 pp.

Newsom, L. D. 1964. Status of the bollworm complex. Beltwide Cotton Prod.-Mech. Conf. Proc, pp. 6-7.

Nirmala Devi, Deshraj and Mahavir Singh 1991. Seasonal abundance of two noctuid pests (*Plusia orichalcea* and *Heliothis armigera*), in northwest Himalayas (India). *J.Ent.* 15(2): 120-124.

Noble, L.W., Glick, P.A. and W.J. Eltel. 1956. Attempts to control certain cotton, corn and vegetable crop insects with light traps. *USDA, Agric. Res.Serv.* 33-28: 3-5.

Oatman, E. R. 1957. Black light traps as a survey tool in apple and cherry orchards. *Ent. Soc. Amer.No. Cent. Br. Proc.* 12: 52-53.

Oatman, E.R. and Brooks, R.F. 1961. Black light- a supplementary survey method for fruit insect populations in Wisconsin. Ent. Soc. Amer. No. Cent. Br. Proc. 16: 118-120.

Odiyo, O.P. 1975. Seasonal distribution and migration of Agrotis ipsilon (Hufn.) (Lepidopera: Noctuidae). *Cent. Overseas Res. Trop. Pest Bull.* 4: 26.

Oh M.S., Lee C.H., Lee S.G., and Lee H.S. 2011. Evaluation of high power light emitting diodes (HPLEDs) and potential attractants for adult *Spodeptera exigua* (Hûbner) (Lepidoptera: Noctuidae). *J Korean Soc Appl Biol Chem* **54**, 416–422.

Oku, T. and Kobayashi, T. 1977. The oriental armyworm outbreak in Tokuku distt. 1960 with special reference to the possibility of mass immigration from China. *Bull.Tokuku Nat. Agri. Expt. Stn.* 55: 105-125.

Oku, T. and Kobayashi, T. 1978. Migratory behaviors and life cycle of noctuids moths with notes on recent status of migrant species in northern Japan. Bull. Tokuku *Nat. Agri. Expt. Stn.* 58: 97-209.

Oman, P. W.1961. What insects are positively photosensitive? *In* Response of Insects to Induced Light, U.S. Dept. Agr., *Agr. Res. Serv.* 20-10, pp. 33-34

Ostmark, H.E. 1968. Bark and ambrosia beetles (Coleóptera: Scolytidae and platypodidae) attracted to an ultraviolet light trap. *Fla. Ent.*51(3): 155-157.

Otake, A. 1966a. Analytical studies of light-trap records in the Hokuriku district. 1. The rice stem borer, *Chilo suppressalis* (Walker) (Lepidoptera: Pyralidae). *Appl. Ent. Zool.* 1 (4): 177-188.

Otake, A. 1966b. Analytical studies of light trap records in the hokuriku district. II. The green rice leaf hopper, *Nephotetix cincticeps. Res. Papul. Ecol.Kyoto univ. Pt.1* pp.63-8.

Otman, E.R. 1964. Orchard insects surveys with black light traps. *J. Econ.Entomol.* 57(1): 6-8.

Otman, E.R. and Longer, E.F. 1962. Fruit insects surveys with black light traps in Wisconsin orchards. *Entomol.Soc. Amer. North Centrl. Br. Proc.* 17: 42-3.

Pandey, V; Sharma, M.K. and R.S. Singh, 2001. Effect of weather parameters on light trap catches of yellow stem borer, *Scirpophaga incertulas* Walker. *Shashpa.* 8: 55-57.

Pandya, H.V.; Shah, A.H. and C.B. Patel. 1993. Influence of light trap on incidence of yellow stem borer in rice. *Gujrat Agril. Univ. Res. J.* 20 (2): 185-187

Pandya, H.V.; Shah, A.H. and M.S. Purohit, 1989. Influence of lunar cycle on light trap catches of rice stem borer, *Scirpophaga incertulas. Oryza.* 26: 108-109.

Parencia, C.R., Jr., Cowan, C.B., Jr., and J.W. Davis, 1962. Relationship of Lepidoptera light-trap collections to cotton field infestations. Jour. Econ. Ent. 55 (5): 692-695.

Parker, J,R., Strand, A.L., and H.L. Seamans, 1921. Pale western cutworm (*Porosagrotis orthogonia* Morr.) *Jour. Agr. Res.* 22 (6): 289-321.

Parrott, P. J. 1927. Progress report on light traps for insect control. Ann. Conv. Empire State Gas and Elec. Assoc, 12 pp. Grand Central Terminal, New York City, Oct. 7.

Parrott, P.J. and Collins, D.L.1935. Some further observations on the influence of artificial light upon codling moth infestations. *J. Econ. Entomol.* 28 (1): 99-103.

Patel, B.S.; Patel, R.C. and M.S. Patel, 1981. A note on the effectiveness of light trap control of Gujrat hairy caterpillar. *G.A.U. Res. J.* 7: 47-48.

Patel, R.C.; Dodia, J.F; Patel, A.G. and D.N. Yadav, 1987. Management of Gujarat hairy caterpillar, *Amsacta moorei* But. by ultra violet light trap. *G.A.U. Res.J.* 12: 59-61.

Patterson, D. F.1936. Four years' experience with electrified light traps. Ontario Ent. Soc. Ann. Rpt.66, pp. 57-61.

Pawar C.S.; Sithanathan S.; Sharma S.C.; Taneja S.L.; Amin P.W.; Lueschner K. and W. Reed 1985. Use and development of insect traps at ICRISAT. Pp: 133-151. In: *Use of traps for pest/vector research and control.* Edt.: Mukhopadhyay, S. and Ghosh, M.R., Bidhan Chandra Krishi Vishwa Vidyalaya, Kalyani, West Bengal.

Pawar,. C.C. Shrivastava C.P. and W. Reed, 1986. Some aspects of population dynamics of *Heliothis armigera* at ICRISAT Center,, III Oriental Entomology

Symposium, Februray 21-24, 1984, Trivandrum. Proceedings March 1986, pp. 79 to 85.

Pedgley, D.E. 1985. Wind born migration of *H. armigera* to the British Isles. *Entomol. Gaz.* 36: 15-20.

Perrot, D.C.F. 1969. A killing agent for use in light traps *N.Z. Entomol.*4(2): 33-8.

Persson, B. 1976. Influence of weather and nocturnal illumination on the activity and abundance of population of Noctuids (Lepidoptera) in South Coastal Queensland. *Bull. Ent, Res.* 66: 33-63.

Peterson, A. and Haeussler, G. J. 1926. The oriental peach moth. *U.S. Dept. Agr. Cir.* 395, 28 pp.

Peterson, A. and Haussler, G.J. 1928. Response of the oriental peach moth and codling moths in coloured lights. *Ann. Entomol. Soc. Amer.*21: 253-379.

Pfrimmer, T. R. 1955. Response of insects to three sources of blacklight. *J. Econ. Entomol.* 48 (5): 619.

Pfrimmer, T. R. 1957. Response of insects to different sources of blacklight. *J. Econ. Entomol.*50 (6): 801-803.

Pfrimmer, T. R. 1961. Light traps to predict need for control. *In* Response of Insects to Induced Light. Sympo. (February 3 and 4, I960.) U.S. Dept. Agr., *Agr. Res. Serv. ARS* 48-50.

Pfrimmer, T. R., Luke fahr, M. J., and J. P. Hollingsworth, 1955. Experiments with light traps for control of the pink bollworm. U.S. Dept. Agr.,*Agr. Res. Serv. ARS*-33-6, March 1955.

Pienkowski, R. L., and J. T. Medler, 1966. Potato leafhopper trapping studies to determine local flight activity. *J. Econ. Entomol.* 59(4): 837-845.

Plaunt, H.N. 1971. Distance of attraction of moths of *Spodoptera littoralis* to BL radiation distances of an ESA. Black light standard trap. *J.Econ. Entomol.* 64(6) 1402-4.

Pol, P. H. Van de.1956. The application of light traps. *Ent. Ber.* 16 (11): 226-236

Popham, E. J. 1960. The uses and limitations of light traps in the study of the ecology of Corixidae (Hemiptera, Heteroptera). *Entomologist* 93 (1167): 162-169.

Porter, L.C. 1941. What kind of light attracts night flying insects most? *Gen. Elect. Rev.* 44: 310-313.

Powers, W. J. 1969. A light-trap bag for collecting live insects. *Jour. Econ. Entomol.* 62 (3): 735-736.

Pradhan, S. 1969. Hairy caterpillar- Control campaign. Chap. 1 pp.16-17. In. Insect pest of crops. NBT *India. Rev.* Edi. 1991.

Pradhan, S. 1973. Physiological adaptations and speciation Chapter 3 In: Insect Phsiology and Anatomy. Edi. Pant and Ghai, ICAR.

Pradhan, S. 1983. Origin of insect pests Chapter 4 in: Agricultural Entomology and pest control. *ICAR Publication, New Delhi.*

Prasad, A.S.R.; Krishnaiah, N.V.; Pasalu, I.C.; Lingaiah, T.; Lakshmi narayanamma, V.; and G. Raju, 2003. Regression models for predicting the peak light trap catches of rice yellow stem borer, *Scirpophaga incertulas* (Walker) based on weather parameters. *Indian J.Pl. Prot.* 31: 34-39.

Pristavka, V. P. 1969. Evaluation of effects of certain abiotic factors on catches of the golden moths (codling moth) by the light traps with ultraviolet radiation. *Jour. Zool. 48* (8): 1177-1184. (In Russian. English Trans.)

Pruitt, D. W.1960. Response of tobacco hornworm moths to ultraviolet radiation. M.S. thesis at Va. Poly. Inst.

Pruthi, H.S. 1969. Text book on Agricultural Entomology, ICAR New Delhi.

Quaintance, A. L., and C. T.Brues, 1905. The cotton bollworm. U.S. Dept. Agr. *Ent. Bull.* 50: 95-96.

Rai, A.B.; Singh, J.; Rai, L. 1990. Rice bug (*Leptocorisa varicornis* Fabr.) - appearance on light trap in eastern Uttar Pradesh, India. *Oryza.,* 27: 66-72.

Rai, A.K.; Khan, M.A. 2002. Light trap catch of rice insect pest, Nephotettix virescens (Distant) and its relation with climatic factors. *Ann.Pl.Prot.Sci.* 10: 17-22.

Rai, A.K; Singh, A.K. and Khan, M.A. 2002. Influence of weather factors on light trap catches of yellow stem borer in *kharif* season. *Indian J. Entomol.* 64: 510-517.

Rai Deepak, Sushil S.N.; Stanley and Veenika Singh 2013, Deployment of Noval Technologies for The Management of White Grubs in Lower Hills of New Himalayan Region, *International Journal of Horticulture,* 3 (2): 3-10.

Raju, S.R. 1959. Response of tobacco hornworm moths to selected narrow wave band ultraviolet energy. M.S. thesis at Va. Poly. Inst.

Rao, B. S. 1964. The use of light traps to control the cockchafer, *Lachnosterna bidentata* (Burmeister) in Malayan rubber plantations. *J. Rubber Res. Inst. Malaya* 18 (5): 243-252.

Rao, B. S. 1965. A light trap for moths *of Nacoleia diemenalis* (Guenee). *J. Econ. Entomol.* 58 (5): 1000-1002.

Reed, W. D., Morrill, A. W., Jr., and E. M. Livingstone, 1935. Trapping experiments for the control of the cigarette beetle. U.S. Dept. Agr.Cir. 356: 14.

Richards, O.W. and Davis, R.G. 1980. A general text book of Entomology. Rev. Edi. Asia Publ. New Delhi.

Riherd, P. T., and Wene, G. P. 1955. A study of moths captured at a light trap at Weslaco, Tex. *Jour. Kans. Ent. Soc.* 28 (3): 102-107.

Riley, C.V. 1892. Attracting by lights. *In* Directions for collecting and preserving insects. *U.S.Nat. Mus. Bul.* 39: 45,51,52.

Riley, C.V. 1885. Light traps for the moths. *In* Fourth report, U.S. Ent. Comn. on the Cotton Worm, pp. 128-131 and 314-321.

Roach, S.H. 1975. *Heliothis zea and H. virusens*: moth activity as measured by black light and pheromone trap. *J.Econ. Entomol.* 68(1): 17-21.

Robinson, H.S. 1952. On the behavior of night flying insects in the neighborhood of a bright light source. *Proc. Royal. London.* (A) 27: 13-21.

Robinson, H.S. and Robinson, P.J.M. 1950. Some notes on the observed behavior of Lepidoptera in flight in the vicinity of light sources tighter with a description of a light trap designed to take entomological samples. *Ent. Gaz.* 1: 03-20.

Robinson, P.J. 1960. An experiment with moths on the effectiveness of a mercury vapour light trap. *Entomol. Gaz.* 11(3): 121-32.

Rohwer,G. and Rohwer, S.A.1964. Test with trap design and killing agent in black light survey traps'. *J. Econ. Entomol.* 57(2): 301-2.

Runner, G. A. 1917. The tobacco beetle and how to prevent damage by it. *U.S. Dept. Agr. F. Bull.* 846: 22

Sandborn, R. 1962. Black light traps - a new tool in peach twig borer control. *West. Fruit Grower* 16 (5): 22.

Sanders, J. G., and S. B. Fracker, 1916. Lachnosterna records in Wisconsin. *J. Econ. Entomol.* 9 (2): 253-261.

Sasamoto, K.M.; Kobayshi and H. Shiraisbi. 1966. Insect control by light trap: I Attracting effectiveness of various lamps of different wave lengths against the green rice leaf hopper, *Nephotetix cincticeps* (Hemiptera: Jassidae) *Jap.J.Appl. Entomol. Zool.* 12 (3): 164-70.

Sathiyanandam, V.K. and R.K.M. Baskaran 1999. Response of groundnut leaf miner *Aproaerema modicella* Deventer to mercury light trap and visible light spectra. *J. Ento. Res.* 23 (3): 209-215

Schaefer, G.W. 1976. Radar observations on insect flight. In: Insect flight. Ed. R.C. Rainey. Symposia of Royal Entomological Society of London Oxford, U.K. Blackwell Publication.

Schuder, D. L. 1964. Attractiveness of black light traps to the European pine shoot moth, *Rhyacionia buoliana* (Schiffermiiller). *N.Cent. Br. Ent. Soc. Amer.* 19: 24-26.

Schuette, V. 1972. Effect of light traps on the population density of *Heliothis zea* (Bodied). *Z. Angew. Entomol.* 70(3): 302-9.

Seamans, H. L., and Gray, H. E. 1934. Design of a new type of light trap to operate at controlled intervals. Quebec. Soc. Protect Plants, 25th and 26th Ann. Rpt., pp. 39-46. Quebec, Canad.

Sen Sharma, P.K. 1985. Insect vectors of sandal spike disease and possible use of light trap for their control. Pp: 103-107. In: *Use of traps for pest/vector research and control.* Edt.: Mukhopadhyay, S. and Ghosh, M.R., Bidhan Chandra Krishi Vishwa Vidyalaya, Kalyani, West Bengal.

Shorey, H.H., and Gaston, L. K. 1965. Sex pheromones of noctuid moths. VIII. Orientation to light by pheromone stimulated males of cabbage looper. *Ann. Ent. Soc. Amer.* 58 (6): 833-836.

Shrivastava, S.K. and Mathur, K.C. 1985. Green leafhopper-relationship of field population and light trap catches. *Oryza.* 22: 240-242.

Slingerland, M. V. 1902. Trap lanterns or "moth catchers." *N.Y. (Cornell) Agr. Expt. Sta. Bul.* 202, 42pp.

Smith, J.S. Jr., Stanley, J. M., Hartsock, J. G., and L. E. Campbell, 1974. S-1 Black-light insect-survey trap. Plans and specifications. U.S.D.A., ARS-S-31, 8 pp.

Smith, J.S., Jr. and Cantelo, W.W.1971. Single vs. multi lamps black light insect trap collection of tobacco horn worm moths. *J. Econ. Entomol. 64(1):* 19-20.

Smith, P.W.1962. The use of black light insect traps as an entomological tool. *Ent. Soc. Amer.* No.Cent. Br. Proc. 17: 38-39.

Smith, P.H., Taylor, J.G. and J.W. Apple. 1960. A comparison of insect traps equipped with 6 and 15 watt. Black light lamps. *J.Econ. Entomol. 52*(6): 1212-14.

Smith, T. A., Boreham, M. M., and K. E. White, 1963. Evaluation of some factors affecting the efficiency of light trap.

Southwood, T.R.E. 1962. Migration of terrestrial arthropods in relation to habitat. *Biol. Rev.* 37: 177-214.

Southwood, T.R.E. 1978. Ecological methods with particular reference to the study of insect populations ELBS and Chapman and Hall Pub. (Revised edi.)

Sparks, A. N.1967. Large-scale field evaluation of electric insect traps to reduce bollworm populations in Reeves County, Tex. U.S. Dept. Agr., Agr. Res. Serv. ARS 33-119,16 pp.

Sparks, A.N., Wright, R.L. and J.P. Hollingsworth. 1967. Evaluation of designs and installations of electric insect trap to collect bollworm moths in Reeves county Texas. *J.Econ.Entomol.60* (4): 929-36.

Squire, F.A. 1943. Phototropism in insects- An Indictment of the light trap method, *Bull. Entomol. Res. 34:* 113-16

Srinavasa, N.; Viraktamath, C.A. and J. Sathyanarayana, 1991. Relative abundance of major insect pests of rice in light trap and their incidence in the field. *Indian J. Entomol.* 53: 603-607.

Srinivasa, N; Viraktamath, C.A. and T.S. Thontadarya, 1990. Influence of moon phase and weather on the light trap catches of insect pests of rice. *Oryza.* 27: 183-190.

Srivastava, V.K.; Diwakar, M.C. and A.D. Pawar, 1992. Light trap and rice pest management. *Plant- Protection- Bulletin*-Faridabad. 44: 39-41.

Stadelbacher, E.A. and T.R. Pfrimmer. 1972. Tobacco bud worms and boll worms: age and mating status of adults collected in light trap in Mississippi. *J.Econ. Entomol 65*(6): 1611.

Stadelbacher,E.A., Laster, M.L. and T.R. Dfrimmer.1972. Seasonal occurrence of populations of Boll worm and Tobacco bud worm moths in the central delta of Mississippi. *Environ. Entomol. 1*(3): 318-23.

Stahl, CF. 1954. Trapping hornworm moths. *J. Econ. Entomol.* 47 (5): 879-882.

Staley, K. A. 1960. Fundamentals of light and lighting. General Electric Co. Bull. LD-2, 95 pp. Cleveland, Ohio.

Stanley, J. M. 1965. Developments in electric insect trap design. [Unpubl.] *Amer. Soc. Agr. Engin.*Paper No. 65-303.

Stanley, J. M. and Dominick, C. B. 1958. Response of tobacco and tomato hornworm moths to black light. *J. Econ.Entomol.* 51 (1): 78-80.

Stanley, J. M. Lawson, F. R., and Gentry, C. R.1964. Area control of tobacco insects with black light radiation. Trans. *Amer. Soc. Agr. Engin.* 7(2): 125-127.

Stanley, J.M. and Dominick, C.B. 1970. Funnel size and lamp wattage Influence on light trap performance. *J.Econ. Entomol.* 63(5): 1423-1426.

Stanley, J.M. and Taylor, E.A. 1965. Population suppression of tobacco hornworms and budworms with black light traps in large-area tests. Conf. Electromagnetic radiation Agr. Proc, pp. 39-41.Publ. by *Eng. Soc. and Agr. Engin.,* Roanoke, Va.

Stanley, J.M., Baumhover, A.H., Smlth, J.S., Jr. Cantelo, W.W., Peace, M.B. and C.A. Asenceio. 1971. A Population suppression experiment for tobacco horn worm and other nocturnal insects using black light traps on an Island, preliminary studies. *USDA ARS.* 42: 193.

Stanley, W.W.1932. Observations on the flight of noctuid moths. *Ann. Ent. Soc. Amer.* 25 (2): 366-368.

Stermer, R. A. 1959. Spectral response of certain stored-product insects to electromagnetic radiation. *J. Econ. Entom.* 52 (5): 888-892.

Stermer, R. A. 1959. Spectral response of certain stored-products insects to electro-magnetic radiation. *J. Econ. Entomol.* 52 (5): 888-892.

Stewart, P. A., Lam, J. J., Jr., and J. D. Hoffman, 1967. Activity of tobacco hornworm and corn earworm moths as determined by traps equipped with black light lamps. *J. Econ. Entomol.* 60 (6): 1520-1522.

Stewart, P.A. 1970. Effect of traps equipped with black light lamps on infestation of lepidopteron larvae in field corn ears'. *J. Econ.Entomol.* 63(6): 1974.

Stewart, P.A. and Lam., Jr. J.J. 1968. Catch of insects at different heights in traps equipped with black light lamp. *Ibid.* 61(5): 1227-30.

Stewart, P.A. and Lam., Jr. J.J. 1969. Hourly and seasonal collection of six harmful insects in traps, equipped with black lamps. *Ibid.* 62 (1): 100-102.

Stewart, P.A., Gentry, C.R., Knott, C.M. and J.J. Lam., Jr. 1968. Seasonal trend in catches of moths of the tobacco horn worm, tomato horn worm and corn ear worm in traps equipped with black light lamps in North Carolina. *Ibid.* 61(1): 43-46.

Stirrett, G. M.1938. A field study of the flight, oviposition, and establishment periods in the lifecycle of the European corn borer, *Pyrausta nubilalis* Hbn. and the physical factors affecting them. II. Flight to light trap. Sei. Agr. 18: 462-484.

Sudia, W. D., and Chamberlain, R. W.1962. Battery-operated light trap, and improved model. Mosq. News. 2: 126-129.New Brunswick, N.J.

Sushil S.N., Mohan M., Bhatt J. C., and H.S. Gupta, 2007, Management of white grubs in Uttarakhand hills, *In National conference on Recent Trends in Rice Pest Management*, CRRI, Cuttak, pp.40

Sushil S.N., Mohan M., Selvakumar G., Bhatt J.C. and H.S. Gupta, 2008, White Grubs of Uttarakhand Hills and their Eco-Friendly Management, *Technical bulletin no. 28*, Vivekananda Parvatiya Krishi Anusandhan Sansthan (Indian Council of Agricultural Research), Almora, pp.49

Sushil S. N.; Pant S.K., and J.C. Bhatt, 2004. Light Trap Catches of White Grub and its Relation with Climatic Factors. *Annals of Plant Protection Sciences*, 12 (2): 254-56.

Sylven, E. 1958. Light trap experiments in studies on fruit leaf Tortricids (Lepidoptera). Swedish State Plant Protect. Inst. Contrib. 11: 74, pp. 212-296.

Tashiro, Haruo, 1961. Relation of physical development and condition of insects to photo sensitivity. *In* Response of Insects to Induced Light. U.S. Dept. Agr., Agr. Res. Serv. ARS 20-10.

Tashiro, Haruo, and Fleming, W. E. 1954. A trap for European chafer surveys. *J. Econ. Entomol.* 47 (4): 618-623.

Tashiro, Haruo, and Tuttle, E. L.1959. Black light as an attractant to European chafer beetles. *J. Econ. Entomol.* 52 (4): 744-746.

Tashiro, Haruo, Hartstock, J.G., and G.G. Rohwer, 1967. Development of black light traps for European chafer surveys. U.S. Dept. Agr.Tech. Bull. 1366, 52 pp.

Tavernetti, J. R., and J. K.Ellsworth, 1938. Energy requirements and safety features of electric insect traps. *Agr. Engin.* 19(11): 485,486,490.

Taylor, J. G., Deay, H. O., and M. T. Orem, 1951. Some engineering aspects of electric traps for insects. *Agr. Engin.* 32 (9): 496,498.

Taylor, J.G. and Deay. H.O. 1950. Electric Lamps and traps in corn borer control. *J.Amer.Soc. Agric. Engin.31*: 503-5.

Taylor, J.G., Altman, L.B., Hollingsworth, J.P., and J.M. Stanley, 1956. Electric insect traps for survey purposes. U.S. Dept. Agr., Agr. Res. Serv. ARS 42-43

Taylor, J.G., Johnson, E.A., Roller, W.L., and H.O. Deay, 1955. Use of electric lamps and traps to protect sweet corn. *In* Purdue Univ. Agr.Expt. Sta. Ann. Rpt. 68: 79.

Taylor, L. R. 1951. An improved suction trap for insects. *Ann. Appl. Biol.* 38 (3): 582-591.

Taylor, L.R. and Brown, E.S. 1972. Effect of light trap design and illumination on samples of moths in the Kenya high lands. *Bull. Entomol. Res.62(1):* 91-112.

Teale, E. W.1961, The insect world of *J. Henri Fabre*. 333 pp. Dodd, Mead, and Co. New York.

Tedders, W. L., Jr., and Osburn, M.1966. Black light traps for timing insecticide control of pecan insects. S.E. Pecan Growers Assoc. 59: 102-104.

Tedders, W.L., Hartstack, J.G., and Osburn, M. 1972. Suppression of hickory shuckworm in a pecan orchard with black light traps. *J. Econ. Entomol.* 65 (1): 148-155.

Thomas, W.W., and Stanley, J.M. 1969. Integrated control as an improved means of reducing populations of tobacco pests. *J. Econ. Entomol.* 62 (2): 1274-1277.

Tietz, H.M. 1936. A novel light trap. *J. Econ.Entomol.* 29(2): 462.

Tiongeo E.R.; Hibino H. and K.C. Ling 1985. The use of light trap and other means in rice tungro studies in the Philippines. pp: 76-86. In: *Use of traps for pest/vector research and control.* Edt.: Mukhopadhyay, S. and Ghosh, M.R., Bidhan Chandra Krishi Vishwa Vidyalaya, Kalyani, West Bengal.

Tiwari, R.K.; Shahi, H.N. and P.R. Singh, 2001. New design of light trap for survey and management of insect pest population of sugarcane based agro-ecosystem. *Indian Journal of Sugarcane Technology* 16: 117-122.

Tomlinson, W. E., Jr. 1970. Effect of black light trap height on catches of moths of three cranberry insects. *J. Econ. Entomol.* 63 (5): 1678-1679.

Tomlinson, W.E., Jr. 1962. The response of crane berry fruit worm to black light. *J.Econ. Entomol. 55* (4): 573.

Treat, A.E. 1962. Comparative moth catches by an ultrasonic and a silent light trap. *Ann. Entomol. Soc. Amer. 55* (6): 716-720.

Tripathi, M.K.; Singh, H.N.; Rakesh Kumar and R. Kumar, 1998. Seasonal abundance and activity of gram pod borer moths, *Helicoverpa armigera* (Hubner) based on light trap catches at Varanasi, Uttar Pradesh. *Environment and Ecology.* 16: 290-293.

Tsao, G. Y.1958. Results on the trapping of moths of pink bollworm and other cotton insects by the black light fluorescent lamp. Knowledge on Insects (Transi, from Chinese) 3: 128-131.

Turner, W. B. 1918. Female *Lepidoptera* at light traps. *Jour. Agr. Res.* 14 (3): 135-149.

Turner, W. B. 1920. *Lepidoptera* at light traps. *Jour. Agr. Res.* 18 (9): 475-481.

U. S. Dept. of Agriculture. 1958. Blacklight insect survey traps aid pest control. USDA News Release 809-58.

Upadhyay, R.N.; Dubey, O.P. and S.M. Vaishampayan, 1999. Studies on the common predatory and parasitic species of insects collected on light trap. *JNKVV. Res. J.* 33: 50-57.

Vail, P.V., Howland, A.F. and T.J. Henneberry. 1968. Seasonal distribution sex ratio, and mating of female noctuid moth in black light trapping studies. *Ann. Entomol. Soc. Amer. 61:* 405-11.

Vaishampayan, S. Jr. and Singh, H.N. 1995. Evidences of the migratory nature of *Heliothis armigera* (Hub.) adults collected on light trap at Varanasi. *Indian J. Ent.* 57: 224-232.

Vaishampayan, S.M. (1979). Final report of ICAR Research Scheme (1975-79) on "Evaluation of various light sources for the control of major Lepidopterous insect pests of economic importance." pp.95, submitted to ICAR, Krishi Bhawan, New Delhi.

Vaishampayan, S.M. 1977. A note on *Helicoverpa armigera* (Hubner) Hardwick, a new generic combination of gram pod borer *Heliothis armigera. Current Sci.46(8): 271.*

Vaishampayan, S.M. 1980. Seasonal abundance and activity of gram pod borer moths *Helicoverpa* (*Heliothis*) *armigera* (Hubner) on light trap equipped with mercury vapour lamp at Jabalpur. *Indian J. Ecol.* **7**: 147-154.

Vaishampayan, S.M. 1981. Light trap studies on Black cutworm *Agrotis ipsilon* (Hfn.) In: *Progress in soil biology and ecology in India.* Edt. Veeresh G.K.: 192-200. *UAS Tech. Sem.* No. 37, Univ. Agril. Sci. Bangalore.

Vaishampayan, S.M. 1982. New design of light trap for survey and management of insect pest population in Agro and Forestry ecosystems. *Indian J. Ent.* 44 (3): 201-205.

Vaishampayan, S.M. 1985a. JNKKV Trap SM 84 Model Light trap for survey of insect pests and vectors. Pp: 152-155. In: *Use of traps for pest/vector research and control.* Edt.: Mukhopadhyay, S. and Ghosh, M.R., Bidhan Chandra Krishi Vishwa Vidyalaya, Kalyani, West Bengal.

Vaishampayan, S.M. 1985b. Factors affecting the light trap catches of insects with emphasis on design aspects. pp: 62-67. In: *"Use of traps for pest/vector research and control".* Edt.: Mukhopadhyay, West Bangal.

Vaishampayan, S.M. 1995. Analysis of factors for the outbreak of polyphagous crop pests creating national problems *Helicoverpa* (*Heliothis*) *armigera,* In: Proceed. *National Seminar on changing pest situation in the current agriculture scenario of India,* ICAR Publication, New Delhi, pp. 261-270.

Vaishampayan, S.M. 2000. Ecology and management systems of *Heliothis* (=*Helicoverpa*) *armigera* Hubner. In: *IPM System in agriculture.* Vol. **7**: 31-54, Aditya Book Pvt. Ltd., New Delhi, India.

Vaishampayan, S.M. 2002. Use of light trap as a component of adult oriented strategy of pest management. In: *Resources management in Plant Protection.* **Vol. II**: 139-144. Publ. Plant Prot. Asso. India, Hyderabad.

Vaishampayan, S.M. 2007. Utility of light trap in integrated pest management. In: *Entomology: Novel Approaches.* Eds: P. C. Jain & M. C. Bhargava.. New India Publishing Agency, New Delhi, pp. 193-210.

Vaishampayan, S.M. and Bahadur A. 1983. Seasonal activity of Teak defoliator *Hyblaea puera* and Teak skeletonizer *Pyrausta machaeralic* monitored by light trap catch In: Insect inter relations in forest and agro ecosystems. Eds. Sen Sharma, Sehgal and Kulshresth: 37-46.

Vaishampayan, S.M. and R. Verma, 1982. Influence of moon light and lunar periodicity on the light trap catches of gram pod borer *Heliothis armigera* (Hubner) moths. *Indian J. Ent.* **44**: 206-212.

Vaishampayan, S.M. and Shrivastava, S.K. 1978. Effect of moon phase and lunar cycle on light trap catch of tobacco caterpillar *Spodoptera litura* Fab. *J. Bombay Nat. Hist. Soc.* 75 (1): 83-87.

Vaishampayan, S.M. and Veda, O.P. 1980. Population dynamics of gram pod borer *Heliocoverpa armigera* (Hubner) and its outbreak situation on gram *Cicer arietinum* L. at Jabalpur. *Indian J. Ent.* **42**(3): 453-459.

Vaishampayan, S.M. and Verma, R. 1983. Comparative efficiency of various light sources in trapping adults of *Heliothis armigera* (Hubn.) *Spodoptera litura* (Boisd) and *Agrotis ipsilon* (Hufn.) (Lepidoptera: Noctuidae). *Indian J. Agric. Sci.* **53** (3): 163-167.

Vaishampayan, S.M. and Verma, R.1987. Seasonal change in reproductive potential of female moths of *Heliothis armigera* (Hubner) (Lepidoptera: Noctuidae) collected on light trap at Jabalpur. *Indian J. Agric. Sci.* **57** (1)

Vaishampayan, S.M.; Verma, R. and K.K. Nema 1984. Reproductive analysis of light trap catches of Teak defoliator *Hyblaea puera* Crammer Lepidoptera: Hyblaedae at Jabalpur, In: Proceed. Third Oriental Entomology symposium Trivendrum Feb.1984, paper No. 107.

Vaishampayan, S.M.; Kogan M.; Waldbauer, G.P. and J.T. Wooley 1975. Spectral specific responses of green house whitefly *Trialeurodes vaporariorum* (Homoptera: Aleyrodaidae). *Ent. Exp. Appl.* 18 (3): 334-356.

Vaishampayan, S.M.; Verma, R. and A.K. Bhoumik, 1987. Possible migration of teak defoliater *Hyblaea puera* (Lepidoptera: Hyblaeidae) in relation to the movement of the south-west monsoon as indicated by light trap catches. *Indian J. Agric. Sci.* **57**: 41-46.

Vaishampayan, S.M.; Verma, R. and K.K. Nema 1984. Seasonal changes in reproductive potential and fat body content of female moths of *Heliothis armigera* (Hubner) collected on light trap at Jabalpur. In: Proceed. First All India Symposium on "Physiology of insect reproduction", deptt. of Zoology, Nagpur University, Nagpur. Dec. 1983 Paper no. 36.

Vallentyne, J. R.1952. Insect removal of nitrogen and phosphorus compounds from lakes. *Ecology,* 33 (4): 573-577.

Vanark, H. and Pretorius, L.M. 1971. Sub sampling of large light trap catches of insects. *Phytophylactica.*3 (1): 29-32.

Veda, O.P. and Vaishampayan, S.M. 1993. Sensitivity of black cutworm *Agrotis ipsilon* (Hufnagel) moths to light. *Indian J.Pl. Prot.* 21: 92-93.

Verheijen, F.J. 1960. The mechanism of trapping effect of artificial light sources upon animals. *Arch. Netherland Zool.* (13): 1-10.

Verma, R.; Vaishampayan, S.M. and R.R. Rawat, 1982. Influence of weather factors on the light trap catches of *Heliothis armigera* (Hubn.) *Indion J. Ent.* 44 (3): 213-218.

Vermorel, V. 1902. The light trap and the destruction of noxious insects. Bur. du "Progress Agricole et viticole." Villefranche et Montpellier. 64 pp.

Wada, Takashi; Kobayashi, T.; Masahiro; Shimazu and Mitsuaki 1980. Seasonal changes in the proportion of mated females in the field population of rice leaf roller *Cnaphalocrosis medinalis* (Lepi.: Pyralidae). *Appli. Ent. Zool.* 15: 81-89.

Wafa, A.K.; Khattab A.A. and F.M. El-Borollosy. 1972. Sex ratio, virgin and gravid female's percentage in *Spodoptera exigua* Hb. population and studies on the use of ultraviolet light traps in reducing the insect population. *Agril. Res. Review,* 50 (1): 37-42.

Wagner, R.E.; Barnes and G.M. Ford. 1969. A battery operated timer and power supply for insect light trap. *J.Econ. Entomol.*62(3): 575-8.

Walkden, H. H., and Whelan, D. B.1942. Owlet moths (Phalaenidae) taken at light traps in Kansas and Nebraska. *U.S.Dept. Agr. Cir.* 643, 26 pp.

Walker, L.T. and Edwards, G.W. 1972. Effect of black light trap design and placement on catch of adult hickory shuckworms. *J.Econ. Entomol.* 65 (6): 1624-27.

Weiss, H.B.1943. Colour perception in insects. *J. Econ. Entomol.* 36 (1): 1-17.

Weiss, H.B.1944. Insect responses to colours. *Jour. N.Y. Ent. Soc.* 52: 267-271.

Weiss, H.B. 1946. Insects and the spectrums. *Jour. N.Y. Ent.Soc.* 54: 17-30.

Weiss, H.B., Soraci, F.A.; and E.E. McCoy, Jr. 1941. Notes on the reactions of certain insects to different wavelengths of light. Jour.N.Y.Ent.Soc. 49: 1-20.

Weiss, H.B.; F.A. Soraci and E.E. McCoy, Jr. 1942. The behavior of certain insects to various wavelength of light. *Jour.N.Y. Ent. Soc.50* (1)1-35.

White, E. G. 1964. A design for the effective killing of insects caught in light traps. *New Zeal. Ent.*3 (3): 25-27.

White, L.D., Hutt, R.B.; Butt, B.A.; Richardson, G.V. and D.A. Backus. 1973. Sterilized codling moths effect of release in a 20 acre apple orchard and comparisons of infestation to trap catches. *Environ. Entomol.2.(5):* 873-80.

Whitehead, F.F. 1946. Codling moth traps in Oklahoma. *J.Econ. Entomol.*39(3): 411.

Wigglesworth, V.B. 1963b. The origin of flights in insects. *Proc. R. Ent. Soc.* London, 28: 23-24.

Wigglesworth,V.B. 1963a. Origin of wings in insects. *Proc. R. Ent. Soc.* London, 197: 97-98.

Wilkinson, R.S. 1966. English entomological methods in the seventeenth and eighteenth centuries. Part I: To 1720. *Ent. Res.* 78: 143-150.

Wilkinson, R.S. 1969. Townsend Glover and the first entomological light trap. *Mich. Entomol.2* (4): 55-62.

Willcocks, F.C. 1916. Insect and related pests of Egypt. Part I. Pink bollworm, trapping the moths of the pink bollworm. *Sultanic Agr. Soc. Cairo.* Vol. 1, pp. 314-315.

Williams, C. B., and Davis, L.1951. Simuliidae attracted at night to a trap using ultraviolet light. *Nature* (London) 179 (4566): 924-925.

Williams, C.B. 1923. A new type of light trap for insects. *Min. Agr. Egypt Tech. Sei. Serv. Bul. 28: 3*

Williams, C.B. 1926. Records from migratory insects chiefly from Africa. *Bull. de. la. Soc. Royale Entomol.* Egypta 10: 224-226

Williams, C.B. 1936. The influence of moon light on activity of certain nocturnal insects, particularly family Noctuidae e indicated by a light trap. Philos. *Trans.R. Soc. Lond.Ser.B.Biol.226*: 256-89.

Williams, C.B. 1937. Butterfly travelers. *Nat. Geo. Mag.* 71: 568-585.

Williams, C.B. 1940. The analysis of four years captures of insects in a light trap. Part 2 the effect of weather conditions on insects activity and the estimation and forecasting of changes in the insect population. *Trans.Roy. Entomol. Soc. London. 90*: 227-306.

Williams, C.B. 1961. Studies on the effect of weather conditions on the activity and abundance of insect populations. *Trans. R. Soc. London Ser. B. Biol. Sci.* 244: 331-78.

Williams, C.B., French, R.A. and M.M. Hosui. 1956. A second experiment on testing the relative efficiency of insect traps. *Bull. Entomol. Res. 46*: 193-204

Williams, C.B., Singh, B.P. and S.E. Ziady. 1956. An investigation in to possible effects of moon light on the activity of insects in the field. Proc. Roy. *Entomol. Soc. Lon.31:* 135-44.

Williams,C.B. 1935. The time of activity of certain nocturnal insects, chiefly *Lepidoptera,* as indicated by a light trap. *Roy. Ent. Soc. London,* Trans. 83 (4): 523-555.

Wimams, C.B. 1939. An analysis of four years' captures of insects in a light trap. Part I. General survey; sex proportions; phenology; and time of flight. *Roy. Ent. Soc. London,* Trans. 89: 79-132.

Wimams, C.B. 1951. Changes in insect populations in the field in relation to proceeding weather conditions. *Proc. Royal Soc. B.* 138: 130-156.

Wolf W.W. 1969. Combined use of sex pheromone and electric traps for cabbage looper control. *Amer.Soc. Agr. Eng.Trans. 12*: 329-35.

Wolf, W. W., Hartstock, J. G., Ford, J. H., and others. 1969. Combined use of sex pheromone and electric traps for cabbage looper control. *Amer. Soc. Agr. Engin.* Trans. 12 (3): 329-331, 335.

Wolf, W.W., Kishaba, A.H., Howland, A.F., and T.J. Henneberry, 1967. Sand as a carrier for synthetic sex pheromone of cabbage loopers used to bait black light and carton traps. *J. Econ. Entomol.* 60 (4): 1182-1184.

Wolf, W.W., Kishaba, A.N., and H.H. Toba, 1971. Proposed method for determining density of traps required to reduce an insect population. Jour. Econ. Entomol. 60 (4): 872-877.

Worthley, H.N. and Nicholas, J.E. 1937. Test with bait and light to trap codling moth. *J. Econ. Entomol.* 30: 417-422

Yadav C.P.S. and Vijayvergia 2000. Integrated management of white grubs in different cropping systems In: IPM System in Agriculture edited by: R.K. Upadhyaya *et al*, Aditya Book Pvt. Ltd. Vol.7 pp.110.

Yadav, L.S.; Chaudhary, J.P. and P.R. Yadav, 1984. Relative abundance of important noctuid moths on light trap infesting chickpea in Haryana. *Bull.Entomol.* 25: 103-110.

Yeh N. and Chung J.P. 2009. High-brightness LEDs-energy efficient lighting sources and their potential in indoor cultivation. *RenewSust Energ Rev* **13**, 2175–2180.

Yoshimeki, M. 1967. A summary of the forecasting programme for rice stem borer in Japan. *The major insect pests of rice plant.* Baltimore, Maryland. The John Hopkins Press, pp.729.

Yothers, M. A. 1928. Are codling moths attracted to lights? *J. Econ. Entomol.* 21: 836-842.

Zheglev, D.T. 1960. Light trap studies of blood sucking dipterans in central Asia. DOKI.ANSSSR, 131: 1430-2.

Zheng Y, Wu W.J., Fu Y.G. 2010. Laboratory evaluation of light-emitting diodes as an attractant for the spiraling whitefly *Aleurodicus dispersus* Russell. *J Environ Entomol* 32(3): 423–426

Zheng, L.X., Zheng, Y. Wu, W.J. and Y.G. Fu. 2014. Field evvalution of different wavelengths Light Emmiting Diodes as attractants for adult *Aleurodicus disperses* Russel (Hemiptera: Aleyrodidae). *Neotrop. Entomol.* 43: 409-414.

Zolkovic, Vladimir, Stancic, Jovan, and Tadic, Milorad. 1958. Results of the application of electric light traps in enticing, detecting and destroying the insects. Nikola Tesla Elec. Res. Inst. 64 pp.

Index

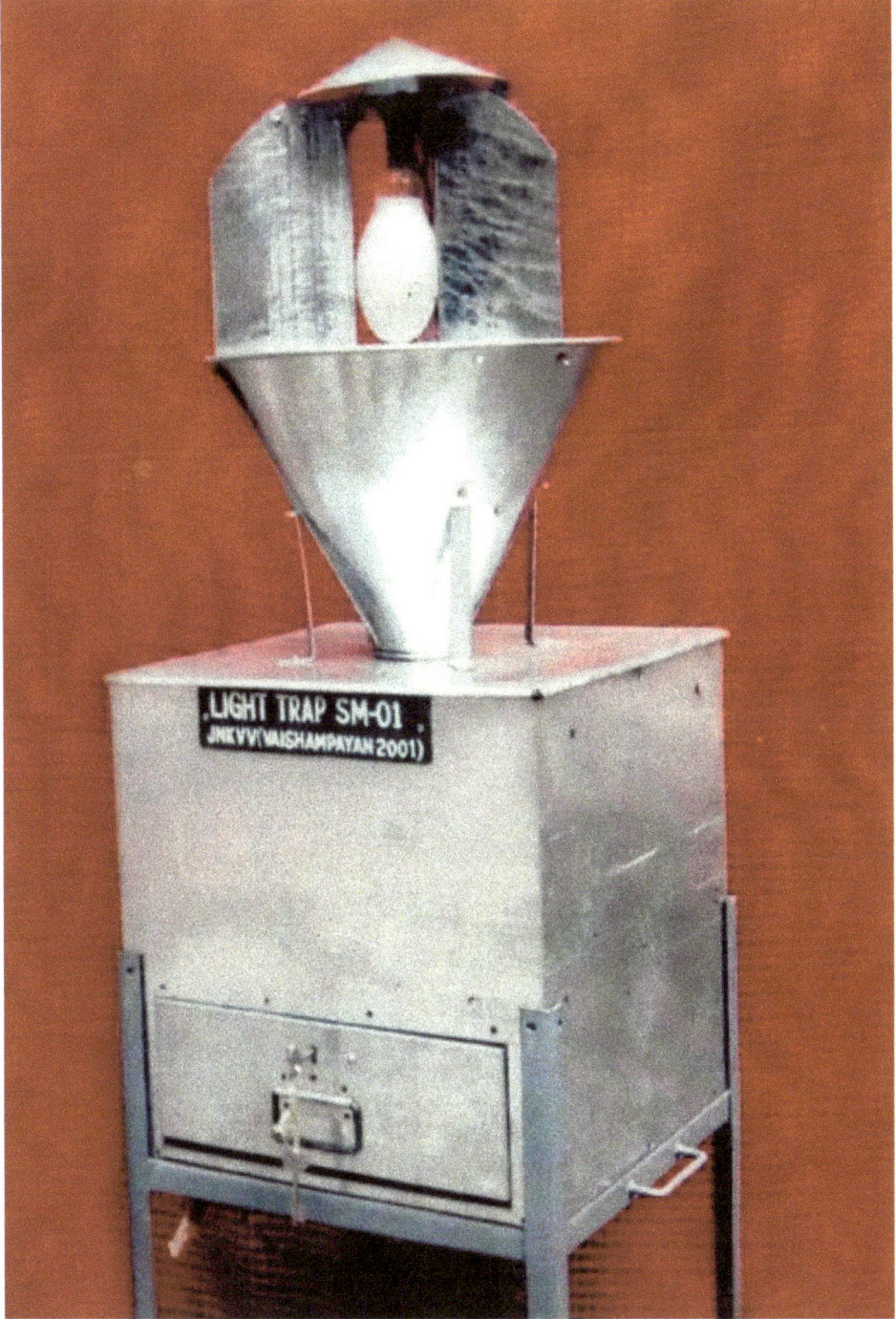

LIGHT TRAP MODEL SM-01

LIGHT TRAP MODEL SM-88

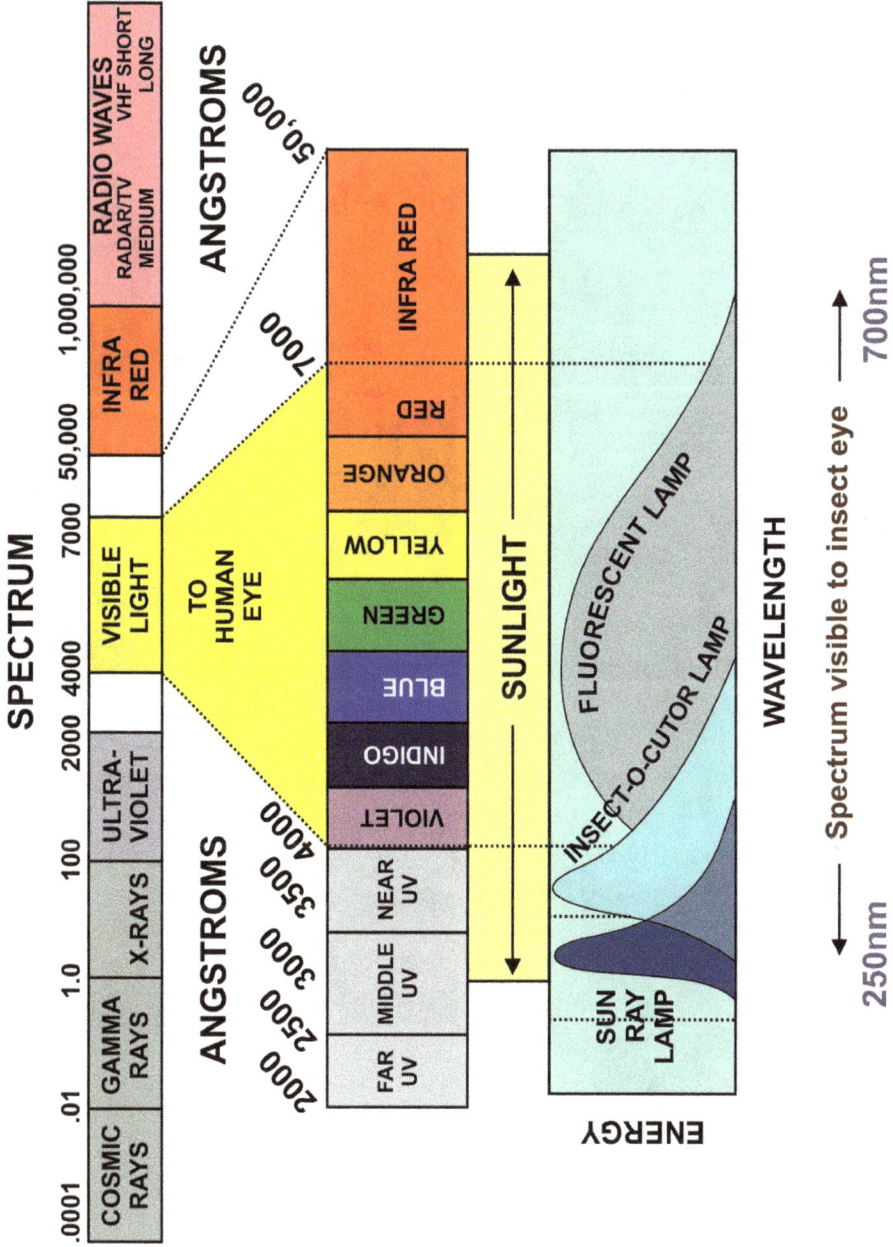

COMPLETE SPECTRUM AND THE PART VISIBLE TO INSECT EYE AND HUMAN EYE

LIGHT TRAP MODEL SM-12

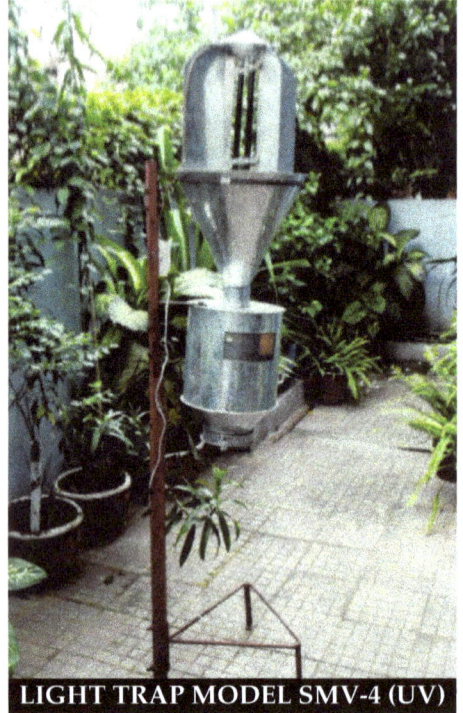
LIGHT TRAP MODEL SMV-4 (UV)

LIGHT TRAP MODEL SMV-3 (UV)